AMERICAN MATHEMATICAL SOCIETY TRANSLATIONS

Series 2

Volume 90

THIRTEEN PAPERS ON FUNCTIONAL ANALYSIS

Z. S. Agranovič
I. A. Bahtin
M. A. Barkar′
M. S. Budjanu
Ja. M. Ceĭtlin
I. C. Gohberg
M. G. Kreĭn
O. V. Kuzel′
S. M. Nikol′skiĭ
G. E. Šilov
A. V. Štraus
S. V. Uspenskiĭ

Published by the
AMERICAN MATHEMATICAL SOCIETY
Providence, Rhode Island
1970

International Standard Book Number 0-8218-1790-6
Library of Congress Catalog Number A51-5559

Printed in the United States of America

TABLE OF CONTENTS

Page

Amer. Math. Soc. Transl.
(2) Vol. 90, 1970

ON SOME QUESTIONS OF ANALYSIS IN HILBERT SPACE. II*

G. E. ŠILOV

§1. Introduction

This article is a continuation of [1]. We give the necessary definitions and results in [1]. Let H be a real, complete, countable-dimensional Hilbert space. By a plane in H we shall mean a finite-dimensional linear manifold (not necessarily containing 0). By a hyperplane in H we shall mean a linear manifold of finite codimension. A hyperplane passing through the point $a \in H$ is designated by a symbol of the form $\alpha(a)$.

Let G be a closed (in the norm) bounded region in H. We call a continuous function in a finite number of arguments of the form $(b_1, x), \cdots, (b_m, x)$ in the region G a *simple* function. A continuous function of the arguments $(b_1, x), \cdots, (b_m, x), |x|^2$ is called a *regular* function. More precisely, a regular function is defined as follows. Fixing the elements $b_1, \cdots, b_m$, we consider the image Q of the region G under the map

$$\mathcal{P}x = [(b_1, x), \cdots, (b_m, x), |x|] \in R_{m+1}.$$

The set Q is bounded along with G. Let the function $\mathcal{F}(\xi_1, \cdots, \xi_m, \eta)$ be defined and uniformly continuous on Q. Then the function $f(x)$ defined on the region G by the equation

$$f(x) = \mathcal{F}[(b_1, x), \cdots, (b_m, x), |x|^2],$$

is called *regular* in G.

In the set $B(G)$ of all regular functions in the region G we introduce the norm

$$\|f\|_B = \sup_{x \in G} |f(x)|.$$

The completion $\overline{B} = \overline{B}(G)$ of the set B in the norm (1) is a ring of functions on G which is closed relative to uniform convergence. In [1] these functions are called the *fundamental* functions. The closure $\overline{\mathfrak{D}}$ of the set $\mathfrak{D} \subset B$ of simple functions is a subring of $\overline{B}$ which coincides with the ring of all weakly continuous functions on G.

---

* Translation of Mat. Issled. 2 (1967), no. 4, 166–186.

If $f(x)$ is a regular function in G, then in any hyperplane $\alpha(a)$ through the point a and orthogonal to the vectors $b_1, \cdots, b_m$ and a the values of $f(x)$ depend only on $|x-a| = r$, so that we may write $f(x) \equiv f_0(r^2)$. The quantity

$$Lf(a) = \lim_{r \to 0} \frac{f_0(r^2) - f(a)}{r^2}$$

(if it exists) is called the *Laplacian of the function* $f(x)$ at $x = a$. If $Lf(x)$ exists at every point $x \in G$ and is also a continuous function of $(b_1, x), \cdots$ $\cdots, (b_m, x)$ and $|x|^2$, then $f(x)$ is called a *smooth regular function.* Completing the set $A(G)$ of smooth functions in the norm

$$\|f\|_A = \sup_{x \in G} |f(x)| + \sup_{x \in G} |Lf(x)|,$$

we obtain a ring of functions $\overline{A} = \overline{A}(G)$ on G, onto which the operator L can be extended by continuity, so that $f_n \longrightarrow f$ (in $\overline{A}$) implies $f_n(x) \longrightarrow f(x)$ and $Lf_n(x) \longrightarrow Lf(x)$ (uniformly in G).

In this paper we take up the problem of giving an intrinsic characterization of functions in the rings $\overline{B}(G)$ and $\overline{A}(G)$ and determining the maximal ideal space of these rings.

§2. Means

We let $C(G)$ designate the complete ring of all (uniformly) continuous functions on the region G.

We choose a basis $e_1, e_2, \cdots$ in the space H and draw a family of planes $p_1 \subset p_2 \subset \cdots$ through a given point $a \in G$, where p_n is parallel to the plane spanned by the basis vectors $e_1, \cdots, e_n$. Let $f(x)$ belong to the ring $C(G)$. The restriction of $f(x)$ to the plane p_n is a continuous function in this plane, and we can construct the mean of this function over any ball and over any sphere with center at a and entirely contained in the region G. We shall let $\mathfrak{M}_n(f, a, r)$ designate the mean of $f(x)$ over the n-dimensional ball $V_n(a, r)$ of radius r with center at a, and we shall let $\mathfrak{M}_n^*(f, a, r)$ designate the mean over the corresponding sphere $S_n(a, r)$. By the mean of the function $f(x)$ over the Hilbert ball $V(a, r)$ of radius r with center at a we mean the number

$$\mathfrak{M}(f, a, r) = \lim_{n \to \infty} \mathfrak{M}_n(f, a, r),$$

if this limit exists. Similarly, by the mean of $f(x)$ over the Hilbert sphere of radius r with center at a we mean the number

$$\mathfrak{M}^*(f, a, r) = \lim_{n\to\infty} \mathfrak{M}_n^*(f, a, r),$$

again under the assumption that this number exists. We note that in this definition the means $\mathfrak{M}^*(f, a, r)$ and $\mathfrak{M}_n^*(f, a, r)$ depend on the choice of basis $e_1, e_2, \cdots$.

To explain the question of existence of means, we consider the following lemmas (compare Lévy [2], Russian p. 228 ff.).

Lemma 1. *Let $S_n^k(a, r, t)$ designate the surface given in the n-dimensional sphere $\{x \in R_n : |x - a| = r\}$ by the inequalities*

$$|(b_1, x - a)| \le t, \cdots, |(b_k, x - a)| \le t,$$

where $b_1, \cdots, b_k$ are fixed elements. Let $S_n(r)$ designate the complete surface of this sphere. Then

$$\lim_{n\to\infty} \frac{S_n^k(a, r, t)}{S_n(r)} = 1.$$

Proof. Without loss of generality, we may assume that $a = 0$. We first consider the case $k = 1$; then, without loss of generality, we may take $|b_1| = 1$.

We have

$$S_n^k(0, r, t) = 2\int_{\phi=0}^{\arcsin\frac{t}{r}} S_{n-1}(r\cos\phi)\, r\, d\phi = 2S_{n-1}(1)\, r^{n-1}\int_0^{\arcsin\frac{t}{r}} \cos^{n-2}\phi\, d\phi,$$

$$S_n(r) = 2\int_0^{\frac{\pi}{2}} S_{n-1}(r\cos\phi)\, r d\phi = 2S_{n-1}(1)\, r^{n-1}\int_0^{\frac{\pi}{2}} \cos^{n-2}\phi\, d\phi.$$

But since for any $\delta > 0$

$$\frac{\int_\delta^{\frac{\pi}{2}} \cos^{n-2}\phi\, d\phi}{\int_0^{\frac{\pi}{2}} \cos^{n-2}\phi\, d\phi} \le \frac{\frac{\pi}{2}\cos^{n-2}\delta}{\frac{1}{2}\,\frac{\Gamma\left(\frac{n-1}{2}\right)\Gamma\left(\frac{1}{2}\right)}{\Gamma\left(\frac{n}{2}\right)}} \le C\sqrt{n}\cos^n\delta \longrightarrow 0,$$

then

$$\lim_{n\to\infty} \frac{S_n^k(0, r, t)}{S_n(r)} = 1,$$

as required.

Hence, most of the surface of the ball is concentrated as $n \longrightarrow \infty$ in an arbitrarily narrow neighborhood of the central cross-section $(b_1, x) = 0$. Since this is also true for any other central cross-section, it is true for any finite intersection of neighborhoods of central cross-sections, which proves Lemma 1.

Lemma 2. *Let $V_n^k(a, r, t)$ designate the volume given in the n-dimensional ball $\{x \in R_n : |x - a| \leq r\}$ by the inequalities*

$$|(b_1, x - a)| \leq t, \cdots, |(b_k, x-a)| \leq t, \; r - t \leq |x - a| \leq r.$$

Let $V_n(r)$ designate the complete volume of this ball. Then

$$\lim_{n \to \infty} \frac{V_n^k(a, r, t)}{V_n(r)} = 1. \tag{1}$$

Proof. For simplicity we take $a = 0$. We first consider the case $k = 1$, $|b_1| = 1$. We have

$$V_n^k(0, r, t)$$

$$= 2 \int_0^{\arcsin \frac{t}{r}} V_{n-1}(r \cos \phi) r \cos \phi \, d\phi - 2 \int_0^{\arcsin \frac{t}{r-t}} V_{n-1}[(r-t)\cos \phi](r-t)\cos \phi \, d\phi$$

$$= 2r^n V_{n-1}(1) \int_0^{\arcsin \frac{t}{r}} \cos^n \phi \, d\phi - 2(r-t)^n V_{n-1}(1) \int_0^{\arcsin \frac{t}{r-t}} \cos^n \phi \, d\phi$$

and

$$V_n(r) = 2r^n V_{n-1}(1) \int_0^{\frac{\pi}{2}} \cos^n \phi \, d\phi .$$

Since

$$\frac{2(r-t)^n V_{n-1}(1) \displaystyle\int_0^{\arcsin \frac{t}{r-t}} \cos^n \phi \, d\phi}{2r^n V_{n-1}(1) \displaystyle\int_0^{\frac{\pi}{2}} \cos^n \phi \, d\phi} \leq \left(1 - \frac{t}{r}\right)^n \longrightarrow 0,$$

and, on the other hand, for any $\delta > 0$, as we saw in Lemma 1,

$$\lim_{n\to\infty} \frac{\int_{\delta}^{\frac{\pi}{2}} \cos^n \phi \, d\phi}{\int_{0}^{\frac{\pi}{2}} \cos^n \phi \, d\phi} = 1,$$

it follows that relation (1) holds for $k = 1$.

Thus most of the volume of the ball is concentrated as $n \longrightarrow \infty$ in an arbitrarily narrow neighborhood of the central cross-section of the sphere where $(b_1, x) = 0$. Since this is also true for any other central cross-section, it is true for the intersection of a finite number of such neighborhoods, which proves Lemma 2.

Theorem 1. *A regular function $f(x)$ has means $\mathfrak{M}(f, a, r)$ and $\mathfrak{M}^*(f, a, r)$ over any ball and sphere contained entirely in the domain of definition G of the function $f(x)$.*

Proof. Let $S(a, r)$ be the sphere of radius r with center at a and contained in the region G, and let $V(a, r)$ be the corresponding ball. We consider a hyperplane $\alpha(a)$ in which the function $f(x)$ depends only on $r = |x - a|$, so that $f(x) = f_0(r^2)$. We show that

$$\mathfrak{M}(f, a, r) = \mathfrak{M}^*(f, a, r) = f_0(r^2).$$

Given $\epsilon > 0$, we find a neighborhood V_ϵ of the sphere $S(a, r) \cap \alpha(a)$ which is interior with respect to the ball $V(a, r)$ and in which the values of the function $f(x)$ differ from $f_0(r^2)$ by less than ϵ. The hyperplane $\alpha(a)$ is defined by equations of the form

$$(b_1, x - a) = 0, \cdots, (b_k, x - a) = 0;$$

the neighborhood V_ϵ can be given by the inequalities

$$|(b_1, x - a)| \leq t, \cdots, |(b_k, x - a)| \leq t, \; r - t \leq |x - a| \leq r.$$

By Lemma 2, for sufficiently large n (we set $n \geq N$)

$$1 - \epsilon \leq \frac{V_n^k(a, r, t)}{V_n(r)} \leq 1.$$

Hence

$$|\mathfrak{M}_n(f, a, r) - f_0(r^2)| = \left| \frac{\int_{V_n(a,r)} f(x)\, dx^n}{V_n(a, r)} - f_0(r^2) \right| =$$

$$= \left| \frac{\displaystyle\int_{V_n^k(a,r,t)} f(x)\,dx^n - \int_{V_n(a,r)-V_n^k(a,r,t)} f(x)\,dx^n}{V_n(a,r)} - f_0(r^2) \right|$$

$$= \left| f_0(r^2)\left[\frac{V_n^k(a,r,t)}{V_n(a,r)} - 1\right] + \frac{1}{V_n(a,r)} \int_{V_n^k(a,r,t)} [f(x)-f_0(r^2)]\,dx^n \right.$$

$$\left. + \frac{1}{V_n(a,r)} \int_{V_n(a,r)-V_n^k(a,r,t)} f(x)\,dx^n \right| \le (|f_0(r^2)| + 1 + |f_0(r^2)| + \epsilon)\epsilon.$$

Since $\epsilon > 0$ is arbitrary, we have $\lim_{n\to\infty} \mathfrak{M}_n(f, a, r) = f_0(r^2)$. Similarly, $\lim_{n\to\infty} \mathfrak{M}^*_n(f, a, r) = f_0(r^2)$, as required.

Using the following lemma, we can extend Theorem 1 to all fundamental functions.

Lemma 3. *If every function* $f_1(x), f_2(x), \cdots$ *has a mean over the ball (sphere)* $S(a, r)$ *and the functions* $f_q(x)$ *converge to* $f(x)$ *as* $q \to \infty$ *uniformly over* $S(a, r)$, *then the function* $f(x)$ *also has a mean over the ball (sphere)* $S(a, r)$, *and*

$$\mathfrak{M}(f, a, r) = \lim_{q\to\infty} \mathfrak{M}(f_q, a, r), \tag{1}$$

$$\mathfrak{M}^*(f, a, r) = \lim_{q\to\infty} \mathfrak{M}^*(f_q, a, r). \tag{2}$$

We go through the proof for the mean $\mathfrak{M}(f, a, r)$. Given $\epsilon > 0$, we find an index q_0 so that for every $q \ge q_0$ we have

$$|f(x) - f_q(x)| \le \epsilon, \quad x \in S(a, r).$$

Hence for any $n = 1, 2, \cdots, q \ge q_0$

$$|\mathfrak{M}_n(f, a, r) - \mathfrak{M}_n(f_q, a, r)| < \epsilon. \tag{3}$$

We further find an index $\mathfrak{N} = \mathfrak{N}(\epsilon, q)$ such that for $n, m \ge \mathfrak{N}$ we have

$$|\mathfrak{M}_n(f_q, a, r) - \mathfrak{M}_m(f_q, a, r)| \le \epsilon. \tag{4}$$

Then, for these values of n and m, (3) and (4) imply

$$|\mathfrak{M}_n(f, a, r) - \mathfrak{M}_m(f, a, r)| \le 3\epsilon.$$

Hence the mean $\mathfrak{M}(f, a, r) = \lim_{n\to\infty} \mathfrak{M}_n(f, a, r)$ exists. Passing to the limit in equation (3) as $n \to \infty$, we obtain

$$\mathfrak{M}(f, a, r) - \mathfrak{M}(f_q, a, r)| \le \epsilon,$$

and hence

$$\mathfrak{M}(f, a, r) = \lim_{q\to\infty} \mathfrak{M}(f_q, a, r),$$

which proves equation (1).

Similar reasoning after replacing $\mathfrak{M}$ by $\mathfrak{M}^*$ gives equation (2).

Thus every function $f(x) \in \bar{B}(G)$ has a mean $\mathfrak{M}(f, a, r)$ over any ball and sphere contained in the region G, and these means coincide. We note that both for functions $f(x) \in B$ and for functions $f(x) \in \bar{B}$ the means do not depend on the choice of basis $e_1, e_2, \cdots$. It would be interesting to show to what extent this last property implies that a given function $f(x) \in C(G)$ belongs to the ring $\bar{B}(G)$.

Example. We consider the function $f(x) - (\Gamma x, x)$, where Γ is a bounded symmetric linear operator. The function $f(x)$ is uniformly continuous in any ball $|x| \le R$, since for $|h| \le 1$

$$|f(x+h) - f(x)| = |(\Gamma(x+h), x+h) - (\Gamma x, x)|$$
$$= |(\Gamma x, h) + (x, \Gamma h) + (\Gamma h, h)| \le \|\Gamma\|(2|x|+1)|h|.$$

We show that *the function* $f(x)$ *has means* $\mathfrak{M}(f, a, r)$ *and* $\mathfrak{M}^*(f, a, r)$ *which do not depend on the choice of basis if and only if the operator* Γ *has the form* $\gamma I + T$, *where* T *is a completely continuous operator.* Since only a linear functional and a constant are added to the function $f(x)$ by translation, it suffices to consider the case $a = 0$.

Let $e_1, e_2, \cdots$ be an orthonormal basis, and let $\|\gamma_{jk}\|$ be the matrix of the operator Γ in this basis. Using the first n vectors of the basis, we compute $\mathfrak{M}_n^*(f, 0, r)$. We have $\mathfrak{M}_n^*(\xi_k^2, 0, r) = (1/n)\mathfrak{M}_n^*(\Sigma_1^n \xi_k^2, 0, r) = r^2/n$ and $\mathfrak{M}_n^*(\xi_j \xi_k, 0, r) = 0$ for $j \ne k$, so that

$$\mathfrak{M}_n^*(f, 0, r) = \frac{r^2}{n} \sum_1^n \gamma_{kk}.$$

Similarly,

$$\mathfrak{M}_n(\xi_k^2, 0, r) = \frac{1}{n}\mathfrak{M}_n\left[\sum_1^n \xi_k^2, 0, r\right] = \frac{1}{nV_n(r)} \int_{V_n(r)} \rho^2 \, d\xi^n$$
$$= \frac{1}{n}\,\frac{r^{n+2}}{n+2}\,\frac{S_n(1)}{r^n V_n(1)} = \frac{r^2}{n+2}, \quad \mathfrak{M}_n(\xi; \xi_k, 0, r) = 0,$$

so that

$$\mathfrak{M}_n(f, 0, r) = \frac{r^2}{n+2} \sum_1^n \gamma_{kk}.$$

Thus, *for given basis* $e_1, e_2, \cdots$ *the means* $\mathfrak{M}(f, 0, r)$ *and* $\mathfrak{M}^*(f, 0, r)$ *exist if and only if the sequence* γ_{kk} *converges in the sense of arithmetic* *means*. We now suppose that the means $\mathfrak{M}(f, 0, r)$ and $\mathfrak{M}^*(f, 0, r)$ exist for any basis $e_1, e_2, \cdots$ and do not depend on the choice of basis. Then the sequence γ_{kk} has an ordinary limit, since otherwise the limit $(1/n)\Sigma_1^n \gamma_{kk}$ would depend on the numbering of the numbers γ_{kk}, i.e. on the order of the basis vectors. Let $\gamma_{kk} \longrightarrow \gamma$ and $T = \Gamma - \gamma I$: the diagonal elements t_{kk} of the matrix T approach zero for any choice of basis. The operator T is symmetric: by von Neumann's theorem [5] there exists a diagonal operator (i.e. an operator having a diagonal matrix in some basis) D such that $D - T = S$ is a Hilbert-Schmidt operator whose matrix $\|\bar{\bar{s}}_{jk}\|$ satisfies the condition $\Sigma_j \Sigma_k s_{jk}^2 < \infty$ in any basis. Let α_k be the diagonal elements of the matrix of the operator D in the basis in which this matrix is diagonal. Since $d_k = t_{kk} + s_{kk}$ and $s_{kk} \longrightarrow 0$, and by assumption we also have $t_{kk} \longrightarrow 0$, it follows that $\alpha_k \longrightarrow 0$, which implies that D is completely continuous. Hence $T = D - S$ is also completely continuous.

Conversely, let $\Gamma = \gamma I + T$, where T is a completely continuous operator. In any basis the relation $t_{kk} \longrightarrow 0$ holds for the matrix $\|t_{jk}\|$ of the operator T, so that, by the above, the means $\mathfrak{M}(f, 0, r)$ and $\mathfrak{M}^*(f, 0, r)$ exist in any basis. It is clear from the proof that in this case

$$\mathfrak{M}(f, 0, r) = \mathfrak{M}^*(f, 0, r) = \gamma r^2.$$

All that has been said about means over a ball and sphere in the Hilbert space H remains true for means over balls and spheres in hyperplanes of the space H. The mean $\mathfrak{M}_\alpha(f, a, r)$ in the hyperplane $\alpha(a)$ over a ball (sphere) of radius r is defined as the limit as $n \longrightarrow \infty$ of the means over the intersection of the n-dimensional ball (sphere) of radius r with the hyperplane $\alpha(a)$. Regular functions $f(x)$ have means $\mathfrak{M}_\alpha(f, a, r)$ in any hyperplane $\alpha(a)$ over any ball with center at a lying in this hyperplane, and they coincide with the corresponding mean over the sphere of the same radius. If we choose the hyperplane $\alpha(a)$ suitably here, it may turn out that the averaging radius r can be taken larger than the distance of the point a from the boundary of the region G. For example, in the case $G = \{x: |x| \le 1\}$, $|a| = 3/5$ the hyperplane $\alpha(a)$ orthogonal to a intersects G in a ball of radius $4/5$, so that the means $\mathfrak{M}_\alpha(f, a, r)$ can be taken over all $r \le 4/5$, while the distance of the point a from the boundary of the ball G only equals $2/5$.

We note that all these means do not depend on the choice of hyperplane $\alpha(a)$: If $\mathfrak{M}_\alpha(f, a, r)$ and $\mathfrak{M}_\beta(f, a, r)$ exist, then $\mathfrak{M}_\alpha(f, a, r) = \mathfrak{M}_\beta(f, a, r)$, since in this case $\mathfrak{M}_{\alpha \cap \beta}(f, a, r)$ exists, and obviously

$$\mathfrak{M}_\alpha(f, a, r) = \mathfrak{M}_{\alpha \cap \beta}(f, a, r) = \mathfrak{M}_\beta(f, a, r).$$

The existence theorem for means in the class of fundamental functions (Theorem 3) completely carries over to means in hyperplanes.

§3. Maximal ideals of the ring $\overline{B}(G)$

The set $\overline{B}(G)$ of all fundamental functions in the region, as the closure with respect to uniform convergence of the ring of regular functions, is again a ring of functions in G with the usual (pointwise) multiplication operation and with the norm

$$\|f\|_{\overline{B}} = \sup_{x \in G} |f(x)|.$$

As with any ring of functions, the values $f(a)$ at points $a \in G$ are multiplicative functionals of the ring $\overline{B}$. But the ring $\overline{B}$ also has other multiplicative functionals. Namely, *if the point a falls in the region G along with a ball $V(a, r)$ lying in the hyperplane $\alpha(a)$, then for $\rho \leq r$ the mean $\mathfrak{M}_{\alpha}(f, a, \rho)$ is a multiplicative functional of the ring $\overline{B}$.* In fact, if $f \in \overline{B}(G)$, $g \in \overline{B}(G)$ and $\beta(a) \subset \alpha(a)$ is a hyperplane in which $f(x)$ and $g(x)$ depend only on $\rho = |x - a|$, so that $f(x) = f_0(\rho^2)$, $g(x) = g_0(\rho^2)$, then in this hyperplane $f(x)\, g(x) = f_0(\rho^2)\, g_0(\rho^2)$ and

$$\mathfrak{M}_{\beta}(fg, a, \rho) = f_0(\rho^2)\, g_0(\rho^2) = \mathfrak{M}_{\beta}(f, a, \rho) \cdot \mathfrak{M}_{\beta}(g, a, \rho).$$

Passing to the limit in the extreme sides of the equation and using Theorem 2, we obtain

$$\mathfrak{M}_{\beta}(fg, a, \rho) = \mathfrak{M}_{\beta}(f, a, \rho) \cdot \mathfrak{M}_{\beta}(g, a, \rho)$$

for all functions $f \in \overline{B}$, $g \in \overline{B}$.

We consider the topological product W of the region G, taken in the weak topology, and the interval $I = \{0 \leq w \leq \sup_{x \in G} |x|^2\}$. Then the multiplicative functional $\mathfrak{M}_{\beta}(f, a, \rho)$ can be associated with the point $\{a, |a|^2 + \rho^2\}$ of the space W. Here the functional $f(a)$ corresponds to the point $\{a, |a|^2\}$. The "upper" subset $N \subset W$ is bounded by the surface $w = |a|^2 + r^2(a)$, where $r(a)$ means the upper bound of the radii averaged at the point a. It is very likely that the set $N = N(G)$ is the entire space of multiplicative functionals of the ring $\overline{B}(G)$, but we can only prove this for a special class of regions G. Namely, we suppose that the region G is characterized by the inequalities

$$|x| \leq \psi((b_1, x), \cdots, (b_n, x)), \qquad (n \leq 0) \tag{1}$$

where the function $\psi(\xi_1, \cdots, \xi_n)$ is defined in a bounded region $Q \subset R_n$, continuous, nonnegative and bounded by the surface

$$|x| = \psi((b_1, x), \cdots, (b_n, x)).$$

Such regions were examined in [1] in connection with solving Dirichlet's problem and inverting the Laplace operator. In the weak topology the region G is a Hausdorff compactum ([1], §7). Through every point a of this region G we can draw a hyperplane $\alpha(a)$ intersecting the region G in a ball with center at a with some radius $r(a)$. The remaining hyperplanes $\beta(a)$ intersect the region G either in a ball of the same radius, or else in a set containing a ball of the same radius.

Theorem 2. *For a region G of the form* (1) *the space $\mathfrak{M}(\overline{B})$ of multiplicative functionals of the ring $\overline{B}(G)$ coincides with the set N on the topological product $W = G \times I$ given by the inequality*

$$|x|^2 \leq w \leq \psi^2((b_1, x), \cdots, (b_n, x)), \ x \in G. \tag{2}$$

Proof. [1] Let $\mathfrak{M}(f)$ be a multiplicative functional of the ring $\overline{B}(G)$. The functional $\mathfrak{M}(f)$ is also multiplicative on the subring $\overline{\mathfrak{D}}(G) \subset \overline{B}(G)$ of all weakly continuous functions on G. Since the region G is a Hausdorff compactum in the weak topology, it follows by a well-known theorem of the theory of normed rings ([3], Russian p. 20) that to the functional $\mathfrak{M}(f)$ there corresponds a point $a \in G$ such that for functions $f(x) \in \overline{\mathfrak{D}}$

$$\mathfrak{M}(f) = f(a).$$

We further let w designate the value of the functional $\mathfrak{M}(f)$ on the function $f(x) = |x|^2$ (which is already not weakly continuous). The number w is not negative and does not exceed the norm of the function $|x|^2$ in the ring $\overline{B}$, i.e. the number $\sup_{x \in G} |x|^2$. Since the weakly continuous functions and $|x|^2$ are generators of the ring $\overline{B}$, it follows that $\mathfrak{M}(\overline{B})$ is included in the topological product of the region G (with the weak topology) and the interval I. We thereby arrive at the above representation. The multiplicative functional $f(a)$ corresponds to the point $\{a, |a|^2\} \in W$, and the mean $\mathfrak{M}_\alpha(f, a, \rho)$ corresponds to the point $\{q, |a|^2 + \rho^2\}$. But we now know that any multiplicative functional $\mathfrak{M}(f)$ corresponds to a point in the set W, and our problem is only to exclude the points $\{\alpha, w\}$, which do not belong to our subset N. Since the space $\mathfrak{M}(\overline{G})$ of multiplicative functionals is the closure of the set $\{x, |x|^2\}$ in the topology of W, it suffices to show that N is closed in W. Suppose that the point $\{a_0, w_0\}$ does not belong to N; it suffices to show

1) Suggested by V. Ja. Lin for the case $G = \{x : |x| \leq 1\}$ (i.e. $n = 0$).

that in the finite-dimensional projection of the set W given by the formula

$$\mathcal{P}\{x, w\} = [(b_1, x), \cdots, (b_n, x), w] \in R_{n+1},$$

the image of the set N is closed and the point $\mathcal{P}\{a_0, w_0\}$ is not in $\mathcal{P}N$. The operator $\mathcal{P}$ takes the set $G \cong [\{x, |x|^2\}]$ to the set $\{(\xi_1, \cdots, \xi_n, w)\}$, where $\xi_k = (b_k, x)$ $(k = 1, \cdots, n)$ and

$$\mu^2(\xi_1, \cdots, \xi_n) \equiv \min_x [|x|^2: (b_k, x) = \xi_k, k = \xi_k, k = 1, \cdots, n] \le w \le \psi^2(\xi_1, \cdots, \xi_n). \qquad (3)$$

We show that $\mathcal{P}G$ fills up this entire set. For fixed $\xi_1 = \xi_1^0, \cdots, \xi_n = \xi_n^0$ the region G contains all points x satisfying the inequality

$$\mu(\xi_1^0, \cdots, \xi_n^0) \le |x| \le \psi((b_1, x), \cdots, (b_n, x)),$$

since on the hyperplane $(b_1, x) = \xi_1^0, \cdots, (b_n, x) = \xi_n^0$ these points form a ball V with center at some point x_0, for which $|x_0| = \mu(\xi_1^0, \cdots, \xi_n^0)$, while on the boundary of this ball $|x| = \psi((b_1, x), \cdots, (b_n, x))$. The operator $\mathcal{P}$ takes this entire ball to the vertical interval $\mu^2(\xi_1, \cdots, \xi_n) \le w \le \psi^2(\xi_1, \cdots, \xi_n)$, filling it up completely, which proves our assertion. In particular, by the assumed continuity of ψ and the obvious continuity of μ, the set $\mathcal{P}G \setminus (3)$ is closed in R_{n+1}.

We show that the points $\{a, |a|^2 + \rho\}$ for $\rho \le r^2(a)$ also go to the set (3). In fact, the number $|a|^2 + r^2(a)$ is exactly the distance of the boundary of the ball V from 0, and the point $\{a, |a|^2 + r^2(a)\}$ maps into the same point in R_{n+1} as any point of the boundary of the ball V. In particular, $|a|^2 + r^2(a) = \psi((b_1, a), \cdots, (b_n, a))$.

If the point $\{a_0, w_0\}$ does not belong to the set N, then either $w_0 < |a_0|^2$, or else $w_0 > \psi^2((b_1, a_0), \cdots, (b_n, a_0))$. The first is easily excluded. We choose an orthonormal basis $e_1, \cdots, e_n, \cdots$. For any m we have $|x|^2 \ge \Sigma_1^m (x, e_k)^2$ so that $w \ge \Sigma_1^m \mathfrak{M}[(x, e_k)^2] = \Sigma_1^m (x, e_k)^2$; as $m \longrightarrow \infty$ we obtain $w \ge |x|^2$. It remains to consider the case $w_0 > \psi^2((b_1, a_0), \cdots \cdots, (b_n, a_0))$. For the projection $\mathcal{P}\{a_0, w_0\}$

$$w_0 > \psi^2(\xi_1, \cdots, \xi_n),$$

so that the point $\mathcal{P}\{a_0, w_0\}$ lies outside the (closed) set $\mathcal{P}N$ and our theorem is proved.

As a symmetric ring with uniform convergence, the ring $\overline{B}$ coincides with the complete ring of continuous functions on the set N. This gives the following intrinsic characterization of functions $f(x) \in \overline{B}(G)$: these functions must have all means $\mathfrak{M}_a(f, a, \rho)$ permitted by the geometry of the region G, and, as functions of the point $\{a, \rho\}$, these means must be continuous in the topology of W.[1)]

§4. Maximal ideals and intrinsic characterization of functions $u(x)$ of the ring $\overline{A}(G)$

As we recall, functions $u(x)$ of the ring $A(G)$ are defined as regular functions on G having a continuous, and hence regular, Laplacian

$$Lu(x) = \lim_{\rho \to 0} \frac{\mathfrak{M}_{\alpha}(u, x, \rho) - u(x)}{\rho^2}. \tag{1}$$

Such functions $u(x)$ are called smooth regular functions. The set $\overline{A}(G)$ is the completion of the ring $A(G)$ in the norm

$$\|u\|_A = \sup_{x \in G} |u(x)| + \sup_{x \in G} |Lu(x)|$$

and is again a ring of functions ([1], §6) on the same set G.

We show that $\overline{A}(G)$ is everywhere dense in $\overline{B}(G)$ in the $\overline{B}$ norm. It suffices to show that every regular function $g(x) = g((a_1, x), \cdots, (a_m, x), |x|^2)$ is the limit with respect to uniform convergence on G of smooth regular functions. We consider the function $g(\xi_1, \cdots, \xi_m, \eta)$ defined and uniformly continuous on the image $Q = \{\mathcal{P}x\} = \{(a_1, x), \cdots, (a_m, x), |x|\}$ of the region G and defining the function $g(x)$ by the rule

$$g(x) = g((a_1, x), \cdots, (a_m, x), |x|^2).$$

All point of the region G lying on the fixed hyperplane $\alpha = \{x \in G: (a_1, x) = \xi_1, \cdots, (a_m, x) = \xi_m\}$ are taken by $\mathcal{P}$ to a fixed "vertical line" in the region Q, taking up there the interval bounded by the numbers

$$\inf_{x \in \alpha} |x|^2 \leq \eta \leq \sup_{x \in \alpha} |x|^2.$$

As we already know, every continuous function $f(\xi_1, \cdots, \xi_m, \eta)$ which is uniformly continuous in Q corresponds to a regular function

$$f(x) \equiv f((a_1, x), \cdots, (a_m, x), |x|^2).$$

If the function $g(x)$ is a smooth regular function, then the corresponding function $g(\xi_1, \cdots, \xi_m, \eta)$ is differentiable with respect to η at every point of the set $\mathcal{P}G$ and its derivative with respect to η is continuous in $\mathcal{P}G$;

---

1) Lin suggested another way of characterizing functions $f(x) \in \overline{B}(G)$, based on the same considerations: if $r(x, y)$ is the metric of the weak topology on G, then functions $f(x) \in \overline{B}(G)$ are precisely those functions on G which are continuous in the metric $r(x, y) + \| |x| - |y| \|$.

and, conversely, if $g(\xi_1, \cdots, \xi_m, \eta)$ is differentiable with respect to η at every point of the set Q and $\partial g/\partial\eta$ is continuous on Q, then $g(x)$ is a smooth regular function. [1)]

Now, having a regular function $g(x)$ and the corresponding uniformly continuous function $g(\xi_1, \cdots, \xi_m, \eta)$ on the set $Q = \mathcal{P}G$, we find a sequence $g_q(\xi_1, \cdots, \xi_m, \eta)$ $(q = 1, 2, \cdots)$ of functions with continuous derivatives which approach $g(\xi_1, \cdots, \xi_m, \eta)$ uniformly on Q. The corresponding functions $g_q(x) \equiv g_q((a_1, x), \cdots, (a_m, x), |x|^2)$ are smooth regular functions in G and converge to $g(x)$ uniformly in G. Hence $A(G)$ is everywhere dense in $B(G)$, and $\overline{A}(G)$ is everywhere dense in $\overline{B}(G)$, as required.

Every multiplicative functional $\mathfrak{M}(f)$ of the ring $\overline{B}(G)$ corresponds to a multiplicative functional $\mathfrak{M}^*(f)$ of the ring $\overline{A}(G)$—namely, the functional $\mathfrak{M}(f)$ itself, considered only on the functions $u(x) \in \overline{A}(G) \subset \overline{B}(G)$. If $\mathfrak{M}_i(f) \neq \mathfrak{M}_2(f)$, then this difference will appear on functions $u(x) \in \overline{A}(G)$, since they are everywhere dense in the ring $\overline{B}(G)$, so that $\mathfrak{M}_1^*(f) \neq \mathfrak{M}_2^*(f)$. But if $m(f)$ is a multiplicative functional of the ring $\overline{A}(G)$, we assert that there is a multiplicative functional $\mathfrak{M}(f)$ of the ring $\overline{B}(G)$ such that $m = \mathfrak{M}^*$. In fact, if we have $m \neq \mathfrak{M}^*$ for some multiplicative functional $\mathfrak{M}(f)$, then there is a function $u_{\mathcal{M}}(x) \in \mathfrak{A}(G)$ in the maximal ideal m and not in $\mathfrak{M}$, i.e. not vanishing at the point $\{a, w\}$, which corresponds to the functional $\mathfrak{M}$. The function $u_{\mathcal{M}}(x)$ can be assumed real and nonnegative (otherwise it can multiplied by its complex conjugate). Since the function $u_{\mathcal{M}}(x)$ does not vanish at the point $\{a, w\}$, it does not vanish in a neighborhood of it; by the usual reasoning using a finite covering, we construct a function $u(x) \in m$ which is positive everywhere on the space $\mathfrak{M}(\overline{B})$ and is again in the ring $A(G)$. Let $u(x) = u((a_1, x), \cdots, (a_m, x, |x|^2)$ where $u(\xi_1, \cdots, \xi_m, \eta)$ is a continuous function in the region $Q \subset R_{m+1}$ and has a continuous derivative with respect to η on the image $\mathcal{P}G \subset R_{m+1}$ of the set G. Since $u(x) \geq c > 0$ on G, we have $u(\xi_1, \cdots, \xi_n, \eta) \geq c > 0$ on $\mathcal{P}G$. But then the function $1/u(\xi_1, \cdots, \xi_n, \eta)$ is continuous on the set $\mathcal{P}G$ and, along with the function $u(\xi_1, \cdots, \xi_n, \eta)$, has a continuous derivative with respect to η there. But then the function $1/u(x)$ is a smooth regular function, i.e. is in the ring $A(G)$; consequently $u(x)$ cannot belong to any maximal ideal of the ring $\overline{A}(G)$. Hence the maximal ideal space $\mathfrak{M}(\overline{A})$ coincides with the

---

1) More precisely, we are always speaking of the existence and continuity of the right derivative; but it is known that continuity of any of the derivatives implies existence of the two-sided derivative (G. E. Šilov, *Mathematical Analysis, Special Course*, GITTL, Moscow, 1961, p. 274).

corresponding space $\mathfrak{M}(\overline{B})$.

In particular, if, as in §3, the region G is characterized by the condition

$$|x| \leq \psi((b_1, x), \cdots, (b_n, x))$$

with suitable function $\psi(\xi_1, \cdots, \xi_n)$, then the space $\mathfrak{M}(\overline{B})$ is known to us: it is given in the topological product $W = G \times I$ by the inequality

$$|x|^2 \leq w \leq \psi^2((b_1, x), \cdots, (b_n, x)), \quad x \in G.$$

But by what has been proved, this set $\mathfrak{M}(\overline{B})$ is the maximal ideal space of the ring $\overline{A}$. The functions $f(x) \in \overline{B}(G)$ correspond to (all) continuous functions $f(x, w)$ on the space $\mathfrak{M}(\overline{B})$; functions $u(x) \in \overline{A}(G)$ correspond to a subring of the ring $\overline{B}(G)$ which is dense in $\overline{B}(G)$. We later give an intrinsic characterization of these functions.

If $u(x) = u((a_1, x), \cdots, (a_m, x), |x|^2)$, the existence of the limit (1) implies the existence of the expression

$$\left.\frac{\partial u((a_1, x), \cdots, (a_m, x), w)}{\partial w}\right|_{1=|x|^2}.$$

We show that the smooth regular function $u(x) = u((a_1, x), \cdots$ $\cdots, (a_m, x), |x|^2)$ has a derivative $\partial u/\partial w$ not only for $w = |x|^2$, but for all w in the interval $|x|^2 \leq w \leq |x|^2 + r^2(x)$, where $r(x)$ is the least upper bound of permissible averaging radii at the point x determined by the geometry of the region G. Let $\alpha(x_0)$ be a hyperplane in which the function $u(x)$ depends only on $\rho = |x - x_0|$ and which contains the intersection of G with the ball $(B\alpha(x_0))$ of radius $r(x_0)$ with center at the point x_0, so that the means $\mathfrak{M}_\alpha(u, x_0, \rho) \equiv u_0(\rho^2)$ are defined for $\rho \leq r(x_0)$. Let $x_1 \in \alpha(x_0)$ and $\rho_1 = |x_1 - x_0| < r(x_0)$. Then at the point x_1 the means $\mathfrak{M}_\alpha(u, x_1, t)$ exist for all $t \leq \sqrt{r^2(x_0) - \rho_1^2}$. Here

$$\mathfrak{M}_\alpha(u, x, t) = u_0(\rho_1^2 + t^2)$$

and

$$Lu(x_1) = \left.\frac{\partial u_0(\rho_1^2 + t^2)}{\partial t^2}\right|_{t=0} = \left.\frac{\partial u_0(\rho^2)}{\partial \rho^2}\right|_{\rho=\rho_1},$$

which proves our assertion.

In addition to the existence of the derivative, we obtain the estimate

$$\left|\left.\frac{\partial u_0(\rho^2)}{\partial \rho^2}\right|_{\rho=\rho_1}\right| \leq \sup_{x \in G}\left|\left.\frac{\partial u((a_1, x), \cdots, (a_m, x), w)}{\partial w}\right|_{w=|x|^2}\right| = \sup_{x \in G}|Lu(x)|. \tag{1}$$

Now suppose that we have a sequence $u_1(x), \cdots, u_n(x), \cdots$ of smooth regular functions which determines a function $u(x) \in \overline{A}(G)$, so that uniformly in G

$$u_n(x) \longrightarrow u(x), \quad Lu_n(x) \longrightarrow Lu(x).$$

The functions $u_n(x)$ correspond to functions $\widetilde{u}_n(x, w)$ on the space $\mathfrak{M}(\overline{A}) = \mathfrak{M}(\overline{B})$ which are continuous and have a continuous derivative with respect to w (in the interval $|x|^2 \leq w \leq \psi^2(x)$). The uniform convergence of the expressions $Lu(x)$ and inequality (1) imply uniform convergence of the derivatives $\partial \widetilde{u}_n(x, w)/\partial w$ on all of $\mathfrak{M}(\overline{A})$. In particular, this convergence is uniform on every interval $|x|^2 \leq w \leq \psi^2(x)$. This implies that the limit function $\widetilde{u}(x, w)$ also has a continuous derivative $\partial \widetilde{u}(x, w)/\partial w$ on every interval $|x|^2 \leq w \leq \psi^2(x)$. This function $\partial \widetilde{u}(x, w)/\partial w$ obviously gives a representation of the regular function $Lu(x)$.

Conversely, let a continuous function $f(x, w)$ be defined on the space $\mathfrak{M}(\overline{B})$ and have a continuous (on all of $\mathfrak{M}(\overline{B})$) derivative $\partial f(x, w)/\partial w$.

We show that the equation $f(x, w) = \widetilde{u}(x, w)$ holds, where $u(x) \in \overline{A}(G)$. In any case, we can write $f(x, w) = \widetilde{u}(x, w)$, where $u(x) \in \overline{B}$, and $\partial f(x, w)/\partial w = \widetilde{g}(x, w)$, where $g(x) \in \overline{B}$. According to [1], §4 there exists a solution of the equation

$$Lv(x) = g(x) \tag{1}$$

with the condition

$$v(z) = u(z), \quad z \in \Gamma, \tag{2}$$

where Γ is the boundary of the region G. The function $v(x)$ is in the ring $\overline{A}(G)$, so that the corresponding function $\widetilde{v}(x, w)$ has a continuous derivative with respect to w; here $\partial \widetilde{v}(x, w)/\partial w$ is the image of the function $Lv(x)$. Since $\widetilde{g}(x, w)$ is the image of the function $g(x)$, we have

$$\frac{\partial \widetilde{v}(x, w)}{\partial w} = \widetilde{g}(x, w) = \frac{\partial f(x, w)}{\partial w} = \frac{\partial \widetilde{u}(x, w)}{\partial w} \tag{3}$$

By assumption, the functions $v(x)$ and $u(x)$ coincide on the boundary Γ of the region G. Hence their means over the spheres of maximum possible radius $r(x)$ coincide at any point $x \in G$, since these sphere lie entirely on the boundary of G. For the corresponding functions $\widetilde{v}(x, w)$ and $\widetilde{u}(x, w)$ we obtain

$$\widetilde{u}(x, \psi(x)) \equiv \widetilde{v}(x, \psi(x)). \tag{4}$$

Integrating equation (3) over w and using initial condition (4), we find

$$\widetilde{u}(x, w) \equiv \widetilde{v}(x, w).$$

Hence $u(x) \equiv v(x) \in \overline{A}(G)$, as asserted.

Thus the function $u(x) \in \overline{B}(G)$ belongs to the ring $\overline{A}(G)$ if and only if the corresponding means $\mathfrak{M}(u, x, \rho) = u_0(\rho^2)$ form a function having a derivative with respect to ρ^2 (for all permissible values of ρ) which is continuous on the set $\mathfrak{M}(\overline{B})$. [1])

Article [4] is a continuation of this paper; in it we solve general differential equations with the operator L.

BIBLIOGRAPHY

[1] G. E. Šilov, *On certain questions of analysis in Hilbert space.* I, Funkcional. Anal. i Priložen. 1 (1967), no. 2, 81–90 = Functional Anal. Appl. 1 (1967), 158–165. MR 35 #4770.

[2] P. Lévy, *Concrete problems of functional analysis,* 2nd ed., Gauthier-Villars, Paris, 1951; Russian transl., "Nauka", Moscow, 1967. MR 12, 834; 36 #6892.

[3] I. M. Gel'fand, D. A. Raĭkov and G. E. Šilov, *Commutative normed rings,* Fizmatgiz, Moscow, 1960; English transl., Chelsea, New York, 1964; 2nd ed., 1967. MR 23 #A1242; 34 #4940.

[4] G. E. Šilov, *On some questions of analysis in Hilbert space.* III, Mat. Sb. 74 (116) (1967), 161–168 = Math. USSR Sb. 3 (1967), 153–158.

[5] a) J. von Neumann, *Characterisierung des Spectrums eines Integraloperators,* Actualités Sci. Indust., no. 229, Hermann, Paris, 1935.

b) G. E. Šilov and Fan Dyk Tin', *Integral, measure and derivative on linear spaces,* "Nauka", Moscow, 1967, pp. 111–113. (Russian) MR 37 #1554.

Translated by:
N. Koblitz

---

1) Or else, applying the above concept of Lin, we can say that it must be continuous on G in the metric $r(x, y) + \|x| - |y\|$, where $r(x, y)$ is the metric of the weak topology in G.

Amer. Math. Soc. Transl.
(2) Vol. 90, 1970

UNCONDITIONAL BASES AND SEMIORDEREDNESS

UDC 519.54

Ja. M. CEǏTLIN

Let E be a real locally convex space with the topology defined by a system of seminorms $\|x\|_{\xi}$, $\xi \in A$. If Σx_i is a convergent series, then various more restricted forms of convergence can be considered for it: 1) reordered convergence, 2) subseries convergence, 3) bounded-multiplier convergence and 4) absolute convergence (for the definitions see [1], page 58 (Russian p.102)). Here 1), 2) and 3) are equivalent if the space is complete with respect to Cauchy sequences, i.e. is semicomplete ([1], page 59 (Russian p. 103)). If E is also a K-space ([2], Russian page 92) in which $|x| \leq |y|$ implies $\|x\|_{\xi} \leq \|y\|_{\xi}$, $\xi \in A$, then E will be called a KT-space. Assuming that E is semicomplete and $\{e_i\}$ is a basis in E, we introduce into E an ordering defined by the cone $K = \{x\colon\ x = \Sigma \alpha_i e_i,\ \alpha_i \geq 0\}$. Under these conditions we have

Proposition 1. *The assertions that* $\{e_i\}$ *is an unconditional basis and that* E *is a* K*-space with respect to the ordering introduced, are equivalent for semicomplete locally convex spaces.*

Proof. If $\{e_i\}$ is an unconditional basis, then, by virtue of the equivalence of reordered and bounded-multiplier convergence, E is in a natural way isomorphic, both as topological vector space and as a lattice, to a normal subspace of the space of all sequences, and hence is a K-space ([2], Russian page 93).

Conversely, if E is a K-space, then it is obviously a discrete K-space ([2], Russian page 132) and, consequently, it is again isomorphic, both as a topological vector space and as a lattice, to a normal subspace of the space of all sequences ([2], Russian page 132). It follows, in view of the equivalence of reordered and bounded-multiplier convergence, that $\{e_i\}$ is an unconditional basis.

In a K-space one can introduce an order topology with the use of the concept of o-convergence ([2], Russian pages 38–41). The following proposition and Theorems 1, 2 and 4 establish that the order topology is not weaker than the original topology for bornological spaces ([3], page 376).

---

*Translation of Izv Vysš. Učebn. Zaved.Matematika 1966, no. 2 (51), 98–104.

Proposition 2. *Suppose E is a semicomplete bornological space and $\{e_i\}$ is an unconditional basis in E. Then a topology in E can be defined by a system of seminorms such that E becomes a KT-space.*

Proof. Let $S(x) = \{y: |y| \le |x|\}$. If $N \subset E$, then $S(N) = \bigcup_{\in N} S(x)$. We define $\|x\|'_\xi = \sup_{y \in S(x)} \|y\|_\xi$, $\xi \in A$, $\|x\|'_\xi < \infty$, $\xi \in A$, for every $x \in E$. Indeed, if $x = \Sigma \alpha_i e_i$ while $y \in S(x)$, then $y = \Sigma \epsilon_i \alpha_i e_i$, $|\epsilon_i| \le 1$. Since $\{e_i\}$ is an unconditional basis, for every $f \in E'$ and every $y \in S(x)$ we have $|f(y)| \le \Sigma |\epsilon_i \alpha_i f(e_i)| \le \Sigma |\alpha_i f(e_i)|$; that is, $S(x)$ is weakly bounded and hence is also strongly bounded. The topology defined by the system of seminorms $\|x\|'_\xi$, $\xi \in A$, is obviously not weaker than the original topology. Since, with respect to the original topology, E is a bornological space, in order to prove that the new topology coincides with the original it is sufficient to show that the stock of bounded sets is the same for both topologies. This is equivalent to the fact that $S(N)$ is bounded in the original topology for every set N bounded in the same topology.

To prove the latter assertion we consider the family of operators $T_{\{\epsilon_i\}}$ defined for all possible $\{\epsilon_i\}$, $|\epsilon_i| \le 1$, $i = 1, 2, \cdots$, in the following manner: $T_{\{\epsilon_i\}}(\Sigma \alpha_i e_i) = \Sigma \epsilon_i \alpha_i e_i$. The operators $T_{\{\epsilon_i\}}$ are continuous. Indeed, a semicomplete bornological space is barreled ([3], page 384) and hence the Banach-Steinhaus Theorem [4] is valid in it. By applying it to the sequence of continuous operators $T^n_{\{\epsilon_i\}} = T_{\{\epsilon_i^1\}}$, $\epsilon_i^1 = \epsilon_i$, $i = 1, 2, \cdots, n$, $\epsilon_i = 0$, $i > n$, we can see the continuity of $T_{\{\epsilon_i\}}$. In order to prove that $S(N) = \bigcup_{\{\epsilon_i\}} T_{\{\epsilon_i\}}(N)$ is bounded for a bounded set N it is sufficient to prove that $\bigcup_{\{\epsilon_i\}} T_{\{\epsilon_i\}} = S(x)$ is bounded for any $x \in E$ [4]. But $S(x)$ is bounded for any x, as was shown above. The proposition is thereby proved.

Corollary. *If E is a semicomplete bornological space with basis $\{e_i\}$ and E_1 is the subspace of E formed by all reordered convergent series of the form $\Sigma \alpha_i e_i$, then either the Banach-Steinhaus Theorem does not hold in E or $E = E_1$, and consequently $\{e_i\}$ is an unconditional basis.*

Proof. E_1 is dense in E. Therefore, if the Banach-Steinhaus Theorem holds, $T_{\{\epsilon_i\}}$ is continuous in E_1 for every $\{\epsilon_i\}$, $|\epsilon_i| \le 1$, and hence is also continuous in E. In that case, by virtue of the equivalence of bounded-multiplier and reordered convergence, $\{e_i\}$ is an unconditional basis.

Theorem 1. *Suppose E is a semicomplete bornological space with an unconditional basis. Then the order topology in E is stronger than the original topology.*

Proof. From Proposition 2 it follows that E can be considered a KT-

space. From the definition of the order topology it is clear that, in order to prove the theorem, we need only show that if $\{x_\beta\}$, $\beta \in \Gamma$, is a direction ([2], Russian page 15) that is decreasing to zero, then $\{x_\beta\}$, $\beta \in \Gamma$, converges to zero in the original topology. Suppose $\{x_\beta\}$, $\beta \in \Gamma$, is a direction that is decreasing to zero, and $x_\beta = \Sigma_i \alpha_i^\beta e_i$, $\alpha_i^\beta \geq 0$, $\beta \in \Gamma$, $i = 1, 2, \cdots$. For any $\xi \in A$, $\epsilon > 0$ and $\beta \in \Gamma$ we have $x_\beta = x_1^\beta + x_2^\beta + \Sigma_{i=1}^n \alpha_i^\beta e_i + \Sigma_{i=n+1}^\infty \alpha_i^\beta e_i$, with $\|x_\beta^2\|_\xi < \epsilon$. Since $\{x_\beta\}$, $\beta \in \Gamma$, decreases to 0 there exists a $\gamma \geq \beta$ such that $|a_i| \leq \epsilon/\|e_i\|_\xi n$, $i = 1, 2, \cdots, n$. Therefore $\|x_\beta\|_\xi \leq \|x_\beta^1\|_\xi + \|x_\beta^2\|_\xi \leq 2\epsilon$ for $\beta \geq \gamma$, i.e. $\{x_\beta\}$ tends to 0 in the original topology.

Theorem 2. *If the conditions of Theorem* 1 *are fulfilled and the order topology is locally convex, then the order topology is equivalent to the original topology.*

Proof. Taking into account Theorem 1 and the fact that E is a bornological space, it is sufficient to prove that the conjugate spaces with respect to the original topology and the order topology are the same. We note that for an order topology the concepts of continuous and o-continuous functionals are equivalent ([3], Russian page 239). For sequences of o-contiuous functionals a theorem of Baluev, which is analogous to the Banach-Steinhaus Theorem, is valid ([3], Russian page 396). Baluev's theorem permits one to describe the space of o-continuous functionals as a space of sequences $\{\phi(e_i)\}$ such that $|\phi(x)| = |\Sigma \alpha_i \phi(e_i)| < \infty$ for every $x = \Sigma \alpha_i e_i$. But each such sequence, by virtue of the Banach-Steinhaus Theorem, determines a functional that is continuous in the original topology, i.e. the space of o-continuous functionals and E' coincide. The theorem is proved.

Let $(a_{\xi,i})$, $\xi \in A$, $i = 1, 2, \cdots$, be an infinite matrix composed of a denumerable number of columns and a certain infinite set of rows, where $0 \leq a_{\xi,i} < \infty$, $\xi \in A$, $i = 1, 2, \cdots$. The space of sequences $\alpha = \{\alpha_i\}$, $i = 1, 2, \cdots$, such that $\|\alpha\|_\xi = \Sigma |\alpha_i| a_{\xi,i} < \infty$ for $\xi \in A$ with a topology given on it by the system of seminorms $\|\alpha\|_\xi$ will be called a Köthe space and denoted by $L(a_{\xi,i})$.

Theorem 3. *If* E *is a semicomplete bornological space with an absolute basis* $\{e_i\}$, *then it is isomorphic to the Köthe space* $L(\|e_i\|_\xi)$.

Proof. We define $\|x\|'_\xi = \|\Sigma \alpha_i e_i\|'_\xi = \Sigma |\alpha_i| \, \|e_i\|_\xi$, $\xi \in A$. Since $\{e_i\}$ is an absolute basis, $\|x\|'_\xi$ is defined for every $x \in E$ and the system of seminorms $\|x\|'_\xi$ defines a certain new topology which is not weaker than the old topology and with respect to which E is a KT-space.

In the same way as in Theorem 1 it can be shown that the new topology is not stronger than the order topology and hence does not have a greater stock of continuous linear functionals. But in the proof of the preceding theorem we established that the conjugate spaces for the original and order topologies coincide, from which it follows that, for the new topology, the conjugate space is preserved, i.e. the stock of bounded sets is preserved, and at the same time also the topology, since E is a bornological space. Theorem 3 was established for Fréchet spaces by Rolewicz [5].*

Theorem 4. *In a locally convex Fréchet space with an unconditional basis the topology definable by a metric and the order topology coincide.*

Proof. Since a Fréchet space is bornological [3′] it follows from Theorem 1 that the order topology is stronger than the topology definable by a metric. On the other hand the topology of a Fréchet space is stronger than the order topology ([3], Russian page 222) since it can be regarded as a KT-space. The proof is complete.

Let E be a linear space of real sequences with the natural ordering and let $e_n = (\underbrace{0, 0 \ldots 0, 1}_{n}, 0, \ldots)$, $e_n \in E$, $n = 1, 2, \cdots$. If $e = \{\alpha_i\} \in E$, then we set $\|e\| = \{|\alpha_i|\}$, where $|e|$ is not necessarily an element of E. We will say that a seminorm p on E is positive monotonic if $0 \leq e \leq e_1$ implies $p(e) \leq p(e_1)$, and that p is monotonic if $|e| \leq |e_1|$ implies $p(e) \leq p(e_1)$. A characteristic of a space with unconditional bases, like sequence spaces, is given by the following.

Theorem 5. *Let E be a linear space of sequences that is semicomplete with respect to a topology definable by a system of monotonic seminorms p_ξ, $\xi \in A$. $\{e_n\}$ is an unconditional basis if and only if the following condition is fulfilled: if $x_n \geq 0$ and $\{x_n\}$ monotonically decreases to zero, then*

$$\lim_{n\to\infty} p_\xi(x_n) = 0, \quad \xi \in A. \tag{1}$$

Proof. If $\{e_n\}$ is an unconditional basis while the topology on E is defined by montonic seminorms, the the fulfillment of (1) is obvious.

Conversely, suppose (1) is fulfilled and $e = \{\alpha_i\}$. Consider $\{e^n\}$, where $e^n = e - \sum_{i=1}^n \alpha_i e_i$. From condition (1) it follows that $\{e^n\}$ converges to zero in the topology, i.e. $e = \sum \alpha_i e_i$. In addition, the functional $f_n(e) = f_n(\{\alpha_i\}) = \alpha_n$ is continuous. Thus the biorthogonal system $\{f_n, e_n\}$ determines a basis. In particular, if E is a Banach space, then the conditions of the theorem are equivalent to the determination of a minimal normed ideal [6].

Editor's note.* The pertinent papers by Rolewicz are: *Remarks on linear metric Montel spaces*, Bull. Acad. Polon. Sci. Sér. Sci. Math. Astr. Phys. **7 (1959), 195–197; *Some remarks on the spaces $N(L)$ and $N(l)$*, Studia Math. **18** (1959), 1–9; MR 21 #7416, 2183.

Let E be a locally convex space with basis $\{e_n\}$. The cone $K = \{x: x = \Sigma\alpha_i e_i, \alpha_i \geq 0\}$ will be called *normal* if the topology of E can be defined by a system of positive monotonic seminorms.

Theorem 6. *Suppose E is a locally convex space with basis $\{e_n\}$ and the cone $K(e_n)$ is normal. Then the following assertions are equivalent:* 1) $E = K - K$ *and* 2) $\{e_n\}$ *is an unconditional basis.*

Proof. From the normality of the cone it follows that for $x \in K$, $x = \Sigma\alpha_i e_i$ the series $\Sigma\alpha_i e_i$ converges reorderedly. Indeed, assuming that the topology of E is defined by monotonic seminorms, we see that, in the completion of E with respect to the topology, the series is subseries convergent. If $E = K - K$, then for every $x = \Sigma\alpha_i e_i$ the series converges reorderedly. The converse is obvious.

Theorem 7. *In order that a basis $\{e_i\}$ in a semicomplete bornological space E be unconditional it is necessary and sufficient that $E = K - K$ and K be a normal cone.*

Theorem 7 follows at once from Proposition 2 and Theorem 6.

Theorem 8. *If E is a semicomplete bornological space and $\{e_n\}$ is a basis in E, then $\{e_n\}$ is unconditional if and only if for each $x = \Sigma\alpha_i e_i$ the series $\Sigma|\alpha_i|e_i$ converges and the following inequality is fulfilled:*

$$\|\Sigma\alpha_i e_i\|_\xi \leqslant M_\xi \|\Sigma|\alpha_i|e_i\|_{\eta(\xi)}, \quad \xi \in A. \tag{2}$$

Proof. The necessity follows from Proposition 2.

Sufficiency. From (2) it follows that, together with $\Sigma\alpha_i e_i$, there converges the series $\Sigma\epsilon_i\alpha_i e_i$, $|\epsilon_i| = 1$. But then for every functional $f \in E'$ the series $\Sigma|\alpha_i f(e_i)|$ converges, and consequently $\{e_n\}$ is an unconditional basis ([7], Proposition 17).

Corollary. *A basis $\{e_n\}$ is unconditional in a semicomplete bornological space if and only if the following inequality is fulfilled:*

$$m_\xi \left\| \sum_{i=1}^n |\alpha_i| e_i \right\|_{\eta(\xi)} \leqslant \left\| \sum_{i=1}^n \alpha_i e_i \right\|_\xi \leqslant M_\xi \left\| \sum_{i=1}^n |\alpha_i| e_i \right\|_{\gamma(\xi)}.$$

For B-spaces, Theorem 8 and the Corollary are proved in [8].

Definition 1. A basis $\{e_n\}$ will be called *boundedly complete* if it follows from $\|\Sigma_{i=1}^n \alpha_i e_i\|_\xi \leq M_\xi$, $\xi \in A$, that the series $\Sigma\alpha_i e_i$ converges.

Definition 2. A KT-space E will be called a KT^**-space* if the following conditions are fulfilled: 1) if $\{x_n\}$ monotonically decreases to zero, then $\{x_n\}$ converges to zero in the topology of E, and 2) if $\{x_n\}$ monotonically increases and is unbounded with respect to the order, with $x_n \geq 0$, $n = 1$,

2, ⋯, then $\|x_n\|_\xi \to \infty$ for some $\xi \in A$.

Definition 3. A K-space E is by definition a *K^*-space* if the boundedness of a set L with respect to the order is equivalent to the fact that, for every numerical sequence $\{\lambda_n\}$ converging to zero, $\{x_n\} \subset L\{\lambda_n x_n\}$ o-converges to zero.

Definition 4. We will say that a locally convex space E possesses the ω-property if from a system of seminorms defining the topology of E one can select a denumerable system $\|x\|_{\xi_n}$, $n = 1, 2, \cdots$, such that $\|x\|_{\xi_n} = 0$, $n = 1, 2, \cdots$, is equivalent to $x = 0$.

We note that every space with a basis possesses the ω-property, inasmuch as one can select $\|x\|_{\xi_n}$ so that $\|e_n\|_{\xi_n} \neq 0$, $i = 1, 2, \cdots$. Certain facts valid for KB-spaces [2] are easily carried over to KT^*-spaces. We introduce the notation $\|x_1, \cdots, x_n\|_\xi = \| |x_1| \vee |x_2| \vee \cdots \vee |x_n| \|_\xi$, $\xi \in A$ (see [2], Russian page 208).

Lemma 1. *In order that* $x_n \overset{o}{\to} 0$ *it is necessary and sufficient that* $\|x_n, \cdots, x_{n+m}\|_\xi \to 0$ *for* $n \to \infty$, $\xi \in A$.

Lemma 2. *If E is a KT^*-space with the ω-property, then the boundedness of $L \subset E$ with respect to the order is equivalent to the inequality* $\|x_1, \cdots, x_n\|_\xi < K_\xi$, $\xi \in A$, *being fulfilled for any finite number of elements* $x_1, \cdots, x_n$, $x_i \in L$, $i = 1, 2, \cdots, n$.

Lemma 3. *Every KT^*-space with the ω-property is a K^*-space.*

The proofs of Lemmas 1–3 are an almost exact repetition of the proofs of the corresponding assertions for KB-spaces ([2], Theorems VII.6.1, VII.6.3 and VII.6.4). Only in the proof of Lemma 2 one must apply a remark ([2], Russian page 223) on the seminorms $\|x\|_{\xi_n}$ determining the ω-property.

Theorem 9. *If E is a semicomplete KT-space in which the ordering is determined by a basis, then the following properties are equivalent:* a) *$\{e_n\}$ is a boundedly complete basis in E,* b) *E is weakly semicomplete,* c) *E is a KT^*-space, and* d) *E is a K^*-space.*

Proof. a) → b). Suppose $\{x_n\}$ is a Cauchy sequence in the weak topology while $\{\phi_i\} \subset E'$ is biorthogonal to $\{e_i\}$; $\phi_i(x_n) \to a_i$, $i = 1, 2, \cdots$. Clearly, $\|\sum_{i=1}^n \alpha_i e_i\|_\xi \leq \sup_n \|x_n\|_\xi$, $\xi \in A$, and since the basis is boundedly complete, $x = \sum_{i=1}^\infty a_i e_i \in E$. The sequence $\{x_n\}$ was taken arbitrarily, and hence E is weakly complete.

b) → c). Condition 1) from the definition of a KT^*-space is always fulfilled for KT-spaces with bases. Therefore one need only prove that condition 2) is fulfilled. Suppose $x_n \geq 0$, $x_n \leq x_{n+1}$, $n = 1, 2, \cdots$. Every functional

$f \in E'$ has the form $f = f_1 - f_2$, $f_1 \geq 0$, $f_2 \geq 0$. If $\{x_n\}$ is bounded in the topology, then $f(x_n)$ is a Cauchy sequence, i.e. $\{x_n\}$ is a Cauchy sequence in the weak topology. It follows from property b) that $\{x_n\}$ is bounded with respect to the ordering, i.e. condition 2) of Definition 2 holds.

c) $\to$ d). This follows from Lemma 3.

d) $\to$ a). Suppose $\{e_i\}$ is not a boundedly complete basis. Then there exists a $\{\lambda_i e_i\} = S$ such that $\Sigma \lambda_i e_i$ diverges, i.e. S is unbounded with respect to the ordering, but $\|\Sigma_{i=1}^n \lambda_i e_i\|_\xi \leq M_\xi$, $\xi \in A$, $n = 1, 2, \cdots$. We note that for an arbitrary sequence $\{\gamma_i\}$ converging to zero the series $\Sigma_{i=1}^\infty \gamma_i \lambda_i e_i$ converges in E. Hence $\{\gamma_n \lambda_n e_n\} \overset{o}{\to} 0$ and therefore E is not a K^*-space. Consequently, if E is a K^*-space, then the basis is boundedly complete.

If E is a K-space, we will denote by $\overline{E}$ the space of o-continuous functionals. We will say that E is *K-reflexive* if $\overline{\overline{E}} = E$ ([2]), Definition IX.6.1).

Theorem 10. *If E is a semicomplete KT-space in which the ordering is defined by a basis, then it is K-reflexive if and only if it is a KT^*-space.*

Proof. It is known ([2], Theorem IX.6.1) that the condition of K-reflexivity is equivalent to the following condition: for every increasing direction $\{x_\alpha\}_{\alpha \in \Gamma}$, $x_\alpha \geq 0$, if $\sup f(x_\alpha) < \infty$ for every $f \in \overline{E}$, $f \geq 0$, then $\sup_{\alpha \in \Gamma} x_\alpha < \infty$. Suppose E is a KT^*-space. We note that in our case $\overline{E}$ is simply the conjugate space E'. Suppose $\{x_\alpha\}_{\alpha \in \Gamma}$ is monotonically increasing and unbounded. Since E is a K^*-space by Theorem 9, there exists a monotonically increasing and unbounded sequence $\{x_{\alpha}\}$ ([2], Theorem VI.6.3). But for this sequence the $\sup_n \phi(x_{\alpha_n}) < \infty$ for every $\phi \geq 0$ cannot be fulfilled, since the fulfillment of this condition is equivalent to $\{x_{\alpha_n}\}$ being a topologically bounded set, and this contradicts the definition of a KT^*-space. Thus E is reflexive.

Conversely, if E is not a KT^*-space, then $\{e_i\}$ is not a boundedly complete basis, and hence the condition of Theorem IX.6.1 in [2] is not fulfilled for some sequence $\{\Sigma_{i=1}^n \alpha_i e_i\}$, $n = 1, 2, \cdots$.

Corollary 1. *If E is a Banach space, then E is a minimal normed ideal and Theorem 10 gives a criterion of reflexivity of a functor defined on the category of Banach spaces satisfying the conditions of Theorem 14a of [6], using the normed ideal E.*

Corollary 2. *Suppose E is defined by the conditions of Theorem 10 and E' is a bornological space with basis $\{f_i\}$, where $\{f_i\}$ is biorthogonal to $\{e_i\}$. Then $\{f_i\}$ is an unconditional boundedly complete basis.*

Proof. $\{f_i\}$ is obviously an unconditional basis. Since E' is a bornological space in the strong topology it can be assumed that E' is a KT-space. It is known that $E' = \overline{E}$ is K-reflexive ([2], Theorem IX.6.2) and hence $\{f_i\}$ is a boundedly complete basis. We note that if E is a bornological space whose ordering is defined by a basis, then E' in the strong topology can also be considered as a KT-space.

Theorem 11. *In order that a semicomplete bornological space E with an unconditional basis $\{e_i\}$ be reflexive it is necessary and sufficient that E and E' be KT^*-spaces.*

Proof. Suppose E is reflexive. Then E is barreled ([9], Proposition 3) and hence the sequence $\{f_i\}$ biorthogonal to the basis is a basis of E' ([3], Proposition 17). But then E' can be regarded as a KT^*-space by virtue of Corollary 2. Since $E = E' = \overline{\overline{E}}$ it follows from Theorem 10 that E is a KT^*-space.

Conversely, suppose E and E' are KT^*-spaces. Then $\{f_i\}$ is a basis of E' by virtue of Theorem 5. From Theorem 10 it follows that $E = \overline{\overline{E}}$ while $\overline{\overline{E}} = E'$. The theorem is proved.

Theorem 11 can be reformulated in the following manner.

Theorem 11$'$. *A semicomplete bornological space with an unconditional basis $\{e_i\}$ is reflexive if and only if $\{e_i\}$ is a boundedly complete basis and the sequence $\{f_i\}$ biorthogonal to $\{e_i\}$ is a basis in E'.*

If E is a Banach space, then Theorem 11 is a special case of Theorem IX.7.4 of [2].

We note in conclusion that all of what has been said can be carried over without difficulty to the case when E is a vector space over the field of complex numbers.

I would like to take the opportunity to express my gratitude to D. A. Raĭkov and B. Z. Vulih, at whose seminars the basic results of this paper were discussed, and to F. S. Vaher and G. Ja. Lozanovskiĭ for many useful discussions.

BIBLIOGRAPHY

[1] M. M. Day, *Normed linear spaces*, Springer-Verlag, Berlin, 1958; Russian transl., IL, Moscow, 1962. MR 20 #1187; 22 #12360.

[2] B. Z. Vulih, *Introduction to the theory of partially ordered spaces*, Fizmatgiz, Moscow, 1961; English transl., Noordhoff, Groningen, 1967. MR 24 #A3494; 37 #121.

[3] L. V. Kantorovič and G. P. Akilov, *Functional analysis in normed spaces*, Fizmatgiz, Moscow, 1959; English transl., Internat. Series of Monographs in Pure and Appl. Math., vol. 46, Macmillan, New York, 1964. MR 22 #9837; 35 #4699.

[3'] L. V. Kantorovič, B. Z. Vulih and A. G. Pinsker, *Functional analysis in partially ordered spaces*, GITTL, Moscow, 1950. (Russian) MR 12,340.

[4] N. Bourbaki, *Espaces vectoriels topologiques*, Actualités Sci. Indust., no. 1189, 1229, 1230, Hermann, Paris, 1955; Russian transl., IL, Moscow, 1959. MR 14, 880; 17, 1109.

[5] Cz. Bessaga and A. Pełczynski, *On the embedding of nuclear spaces in the space of all infinitely differentiable functions on the line*, Dokl. Akad.. Nauk SSSR 134 (1960), 745–748 = Soviet Math. Dokl. 1 (1960), 1122–1125. MR 24 #A415.

[6] B. S. Mitjagin and A. S. Švarc, *Functors in categories of Banach spaces*, Usephi Mat. Nauk 19 (1964), no. 2 (116), 65–130 = Russian Math. Surveys 19 (1964), no. 2, 65–127. MR 29 #3866.

[7] J. Dieudonné, *On biorthogonal systems*, Michigan Math. J. 2 (1954), 7–20. MR 16, 47.

[8] B. E. Veĭc, *Some characteristic properties of unconditional bases*, Dokl. Akad. Nauk SSSR 155 (1964), 509–512 = Soviet Math. Dokl. 5 (1964), 436–439. MR 29 #3865.

[9] N. Bourbaki, *Sur certains espaces vectoriels topologiques*, Ann. Inst. Fourier Grenoble 2 (1950), 5–16. MR 13, 137.

Translated by:
S. Smith

Amer. Math. Soc. Transl.
(2) Vol. 90, 1970

IMBEDDING THEOREMS FOR FUNCTIONS WITH PARTIAL DERIVATIVES CONSIDERED IN VARIOUS METRICS*

S. M. NIKOL'SKIĬ

In this article imbedding theorems are extended to a more general class of functions of several variables whose partial derivatives with respect to different variables are considered in different metrics.

§1

1.1. In articles [10, 11, 12] we have generalized the imbedding theorems of S. L. Sobolev [16, 17] and of V. I. Kondrašov [9, 17] for the function classes $H_p^{(r_1, \cdots, r_n)}$. The functions in these classes may have nonmixed partial derivatives of different orders with respect to the individual variables $x_1, \cdots, x_n$. But all the partial derivatives and the Hölder conditions that they must satisfy are considered only in the metric of L_p. However, it is sometimes necessary to deal with functions f of several variables whose partial derivatives with respect to different variables are considered in different metrics. For example, we may speak of functions $f(x_1, \cdots, x_n)$, integrable (continuous) on a domain G and such that the integral

$$I(f) = \int_G \sum_{i=1}^{n} \left| \frac{\partial^{r_i} f}{\partial x_i^{r_i}} \right|^{p_i} dG < M < \infty, \tag{1}$$

makes sense, where the r_i are given natural numbers and $1 \leq p_i \leq \infty$ $(i = 1, \cdots, n)$. We then raise the question of which partial derivatives of f will be integrable in the metric of L_q. In particular this is necessary in order to clear up whether the function f will satisfy a Lipschitz condition of some degree (in the metric of C). A positive solution to this problem gives an equicontinuity condition for a bounded class of functions for which (1) holds.

1.2. In this article we will introduce a new class of functions of several variables $H_{p_1, \cdots, p_n}^{(r_1, \cdots, r_n)}(G; M)$, determined, in general, by various numbers $r_i > 0$ and p_i, where $1 \leq p_i \leq \infty$. It is defined as the intersection of the

---

* Translation of Izv. Akad. Nauk SSSR Ser. Mat. 22 (1958), 321–336.

classes $H^{(r_i)}_{p_i x_i}(G; M)$, $i = 1, \cdots, n$.[1])

Thus for functions $f \in H^{(r_1, \cdots, r_n)}_{p_1, \cdots, p_n}$ the partial derivatives with respect to the individual variables x_i are considered in the various metrics of the spaces L_{p_i}.

In §2 we prove an imbedding theorem (§2.4) for the class $H^{(r_1, \cdots, r_n)}_{p_1, \cdots, p_n}$, when $G = R_n$ is the entire n-dimensional space. The proof is carried out by approximating the function f by entire functions of finite order (exponential type). Here it is necessary to generalize Jackson's theorem in connection with the classes $H^{(r_1, \cdots, r_n)}_{p_1, \cdots, p_n}$. The essential apparatus in this proof consists of two inequalities for entire functions $g = g_{\nu_1, \cdots, \nu_n}$ of orders $\nu_1, \cdots, \nu_n$ corresponding to the variables $x_1, \cdots, x_n$:[2])

$$\|g\|_{L^{(n)}_{p'}} \leqslant 2^n \left(\prod_1^n \nu_i\right)^{\frac{1}{p} - \frac{1}{p'}} \|g\|_{L^{(n)}_p} \quad (1 \leqslant p < p' \leqslant \infty), \qquad (1)$$

$$\|g\|_{L^{(m)}_p} \leqslant 2^n \left(\prod_{m+1}^n \nu_i\right)^{\frac{1}{p}} \|g\|_{L^{(n)}_p} \quad (1 \leqslant p \leqslant \infty, \quad 1 \leqslant m < n), \qquad (2)$$

where

$$\|f\|_{L^{(m)}_p} = \left(\int_{-\infty}^{\infty} \cdots \int_{-\infty}^{\infty} |f(x_1, \ldots, x_m, x_{m+1}, \ldots, x_n)|^p dx_1 \ldots dx_m\right)^{\frac{1}{p}}.$$

These and analogous inequalities for trigonometric polynomials were obtained by us in [10, 11]. In the case of trigonometric polynomials inequality (1) reduces to an inequality obtained by Jackson [6] for $n = 1$ and $p' = \infty$.

---

1) The classes $H^{(r)}_{px_i}$ were introduced in our articles [11, 12]. Let $r > 0$, $r = \bar{r} + \alpha$, where $\bar{r}$ is an integer and $0 < \alpha \leq 1$. A function $f(P)$, defined on G, belongs to the class $H^{(r)}_{px_i(G)}(M)$, if it and its partial derivatives up to order $\bar{r}$ are p-integrable on G and the partial derivative $f^{(\bar{r})}_{x_i}$ satisfies the inequality

$$\| f^{(\bar{r})}_{x_i}(\bar{P} + \bar{h}) - f^{(\bar{r})}_{x_i}(P)\|_{L_p(G')} \leqslant M |\bar{h}|^{\alpha},$$

if $0 < \alpha < 1$, and the inequality

$$\| f^{(\bar{r})}_{x_i}(\bar{P} + \bar{h}) - 2f^{(\bar{r})}_{x_i}(P) + f^{(\bar{r})}_{x_i}(P - \bar{h}) \|_{L_p(G')} \leqslant M |\bar{h}|,$$

if $\alpha = 1$. Here G' is any domain, $G' \subset G$ and $\bar{h}$ is any vector parallel to the x_1-axis such that the vectors $\pm \bar{h}$ translate G' within the limits of G. If $G = R_n$, then we may take $G' = R_n$ in the definition. In this case we set $H^{(r)}_{px_1(G)} = H^{(r)}_{px_1}$.

2) The constant 2^n can be improved, but the most precise value of the constant in the general case is still unknown (see the works of N.K. Bari [3] and I. I. Ibragimov [8], containing results related to this problem).

We give an example in §2.6 that shows, under well-known restrictions (see condition (1) of that subsection) which actually are thought to be inessential, that the theorem that we prove cannot be strengthened, at least in terms that apply here.

In §3 we state a compactness theorem for the classes $H^{(r_1, \cdots, r_n)}_{p_1, \cdots, p_n}$ analagous to the corresponding theorem for the classes $H^{(r_1, \cdots, r_n)}_{p}$ [14].

In §4 we show that we can obtain analogous imbedding theorems in other cases (for certain other domains G; for the periodic case) from Theorem 2.4, using a theorem on the continuation of functions, or, analogously, as a corollary.

In particular, it follows from these results (see §4.2) that the set of functions $f(x_1, \cdots, x_n)$ bounded by a given constant, and defined on a domain G which is a parallelepiped $a_i \leq x_i \leq b_i$ and such that the inequality of §1.1 holds, will be equicontinuous if

$$1 - \sum_{1}^{n} \frac{1}{p_i r_i} > 0.$$

For such a domain G this assertion contains results of E. Gagliardo [7, Theorems 1.I, 1.II, 1.III].

In conclusion I would like to note that this article in some measure answers a question raised by S. L. Sobolev: can the methods worked out in my articles on the theory of differentiable functions of several variables be used to obtain imbedding theorems, taking into account that different partial derivatives of the functions are considered in different metrics?

§2

2.1. Let the numbers r_i, p_i, $M_i (i = 1, 2, \cdots, n)$ satisfy the inequalities

$$r_i > 0, \quad 1 \leqslant p_i \leqslant \infty.$$

Thus, in particular, some of the p_i may be infinite.

By definition, a function $f = f(x_1, \cdots, x_n)$ of n variables, defined on the n-dimensional space R_n of points $(x_1, \cdots, x_n)$, belongs to the class $H^{(r_1, \cdots, r_n)}_{p_1, \cdots, p_n}(M_1, \cdots, M_n)$, if it belongs to $H^{(r_i)}_{p_i x_i}(M_i)$ with respect to each variable x_i. If $M_1 = \cdots = M_n = M$ we will write $H^{(r_1, \cdots, r_n)}_{p_1, \cdots, p_n}(M)$.

Let $g_{\nu_1, \cdots, \nu_n} = g_{\nu_1, \cdots, \nu_n}(x_1, \cdots, x_n)$ denote an entire function of orders $\nu_1, \cdots, \nu_n$ in the variables $x_1, \cdots, x_n$, respectively. This definition makes sense when all the ν_i $(i = 1, \cdots, n)$ are finite and nonnegative. We extend this definition to the case when some but not all of the ν_i are infinite. Namely, if

$$\nu_{n_1} = \nu_{n_2} = \ldots = \nu_{n_m} = \infty \quad (m < n),$$

and the remaining $\nu_k < \infty$, then this will mean [5] that for almost all $(x_{n_1}, \cdots, x_{n_m})$ in the corresponding m-dimensional coordinate subspace, the function $g_{\nu_1, \cdots, \nu_n}$ is an entire function in each of the remaining variables of corresponding order. For example, $g_{\nu_1, \nu_2, \infty, \cdots, \infty}$ means that this function, for almost all $(x_3, \cdots, x_n)$ in the sense of $(n-2)$-dimensional measure is an entire function in x_1, x_2 of order ν_1, ν_2, respectively.

2.2. Theorem (Generalized Jackson Inequality). *There exists a constant c, depending on r_i and p_i $(i = 1, \cdots, n)$, such that for each function $f \in H^{(r_1, \cdots, r_n)}_{p_1, \cdots, p_n}(M_1, \cdots, M_n)$, there is a system of functions*

$$g_{\nu_1, \ldots, \nu_n} \quad (1 \leqslant \nu_i \leqslant \infty, \quad i = 1, \ldots, n), \tag{1}$$

for which

$$\left.\begin{aligned} &\| f - g_{\nu_1, \infty, \ldots, \infty} \|_{L_{p_1}^{(n)}} < \frac{cM_1}{\nu_1^{r_1}}, \\ &\| g_{\nu_1, \infty, \ldots, \infty} - g_{\nu_1, \nu_2, \infty, \ldots, \infty} \|_{L_{p_2}^{(n)}} < \frac{cM_2}{\nu_2^{r_2}}, \\ &\cdots\cdots\cdots\cdots\cdots\cdots \\ &\| g_{\nu_1, \ldots, \nu_{n-1}, \infty} - g_{\nu_1, \ldots, \nu_n} \|_{L_{p_n}^{(n)}} < \frac{cM_n}{\nu_n^{r_n}}. \end{aligned}\right\} \tag{2}$$

Similar inequalities may be obtained by rearranging the variables $x_1, \cdots, x_n$ (and $\nu_1, \cdots, \nu_n$ respectively). In (2) *the letters ν_i denote finite numbers* (≥ 1).

Moreover,

$$\| g_{\nu_1, \ldots, \nu_n} \|_{L_{p_i}^{(n)}} \leqslant c \| f \|_{L_{p_i}^{(n)}}. \tag{3}$$

The constant c is independent of $\nu_i = 1$ $(i = 1, \cdots, n)$.

Remark. For $p_1 = \cdots = p_n = p$ the inequalities (2) imply the well-known inequality

$$\| f - g_{\nu_1, \ldots, \nu_n} \|_{L_p} < c \sum_1^n \frac{M_k}{\nu_k^{r_k}} \tag{4}$$

[11, Theorem 7] (for $p = \infty$ and $n = 1$ see [4]), a generalization of the Jackson inequality.

The proof of the theorem is similar to the proof of (4) (see [11]; for

$p=\infty$ see [4]). Namely, we introduce entire functions $K_\nu^{(r)}(\mu)$ of orders ν $(1 \le \nu < \infty,\ r>0)$, such that

$$\int_{-\infty}^{\infty} K_\nu^{(r)}(u)\,du = 1, \tag{5}$$

$$\int_{-\infty}^{\infty} |K_\nu^{(r)}(u)|\,du = a_r < \infty, \tag{6}$$

and

$$\left\| f - \int_{-\infty}^{\infty} K_\nu^{(r)}(u-x_1) f(u, x_2, \ldots, x_n) \right\|_{L_p} < \frac{b_r M}{\nu^r} \quad (\nu \geqslant 1,\ 1 \leqslant p \leqslant \infty) \tag{7}$$

for all $f \in H_{px_1}^{(r)}(M)$ (here b_r is independent of p).

Take, for example, $n=3$; then set

$$\left.\begin{aligned} g_{\nu_1,\infty,\infty} &= \int_{-\infty}^{\infty} K_{\nu_1}^{(r_1)}(u) f(x_1+u, x_2, x_3)\,du, \\ g_{\nu_1,\nu_2,\infty} &= \int_{-\infty}^{\infty}\int_{-\infty}^{\infty} K_{\nu_1}^{(r_1)}(u_1) K_{\nu_2}^{(r_2)}(u_2) f(x_1+u_1, x_2+u_2, x_3)\,du_1\,du_2, \\ g_{\nu_1,\nu_2,\nu_3} &= \int_{-\infty}^{\infty}\int_{-\infty}^{\infty}\int_{-\infty}^{\infty} K_{\nu_1}^{(r_1)}(u_1) K_{\nu_2}^{(r_2)}(u_2) K_{\nu_3}^{(r_3)}(u_3)\,\cdot \\ &\quad \cdot f(x_1+u_1, x_2+u_2, x_3+u_3)\,du_1 du_2\,du_3. \end{aligned}\right\} \tag{8}$$

The inequalities (2) are simple corollaries of properties (5), (6) and (7) of the kernel $K_\nu^{(r)}$. For example, the third inequality emerges from the following computations:

$$\begin{aligned} &(g_{\nu_1,\nu_2,\infty} - g_{\nu_1,\nu_2,\nu_3})(x_1, x_2, x_3) \\ &= \int_{-\infty}^{\infty}\int_{-\infty}^{\infty} K_{\nu_1}^{(r_1)}(u_1) K_{\nu_2}^{(r_2)}(u_2) h(x_1+u_1, x_2+u_2, x_3)\,du_1 du_2, \end{aligned} \tag{9}$$

where

$$h(v_1, v_2, v_3) = f(v_1, v_2, v_3) - \int_{-\infty}^{\infty} K_{\nu_3}^{(r_3)}(u) f(v_1, v_2, v_3+u)\,du.$$

Therefore, applying the generalized Minkowski inequality to (9) and applying (5) and (7) we obtain

$$\| g_{\nu_1,\nu_2,\infty} - g_{\nu_1,\nu_2,\nu_3} \|_{L_{p_3}^{(n)}} \leqslant a_{r_1} a_{r_2} \| h \|_{L_{p_3}^{(n)}} \leqslant \frac{a_{r_1} a_{r_2} b_{r_3}}{\nu_3^{r_3}} \quad (\nu \geqslant 1).$$

Finally we will show, for the case $n = 3$, how (3) is proved. For this it is sufficient to apply the generalized Minkowski inequality to the integral (8), which gives

$$\| g_{\nu_1,\nu_2,\nu_3} \|_{L_{p_i}^{(n)}} \leqslant a_{r_1} a_{r_2} a_{r_3} \| f \|_{L_{p_i}^{(n)}} \qquad (i = 1,\ 2,\ 3)$$

i.e. we obtain (3) for $n = 3$.

In the sequel we will use the following lemma, which generalizes the converse of the Bernštein approximation theorem.

2.3. Lemma. *Let $r > 0$, $f \in L_p^{(n)}$ and suppose the inequality*

$$\| f - g_{\nu,\infty,\ldots,\infty} \|_{L_p^{(n)}} < \frac{K}{\nu^r}$$

holds for a sequence $g_{\nu,\infty,\cdots,\infty}$, where $\nu = a^s$ is a geometric progression ($s = 0, 1, \cdots$; $a > 1$). Then $f \in H_{px1}^{(r)}(M)$, where

$$M < c\,(\| f \|_{L_p^{(n)}} + K)$$

and the constant c is independent of the adjacent factor.

See Theorem 8 in [11] for the proof (the statement and remark are at the end of the proof of that theorem).

2.4. Theorem. *Let $r_i > 0$, $1 \le p_i \le q \le \infty$, let n and m be natural numbers, $1 \le m \le n$, and let*

$$\rho_i = \frac{r_i \varkappa}{\varkappa_i} > 0 \quad (i = 1, \ldots, n), \tag{1}$$

where

$$\varkappa = \begin{vmatrix} 1 - \sum_1^n \dfrac{\frac{1}{p_l} - \frac{1}{q}}{r_l} & -\dfrac{1}{q}\sum_1^n \dfrac{1}{r_l} \\ -\sum_{m+1}^n \dfrac{\frac{1}{p_l} - \frac{1}{q}}{r_l} & 1 - \dfrac{1}{q}\sum_{m+1}^n \dfrac{1}{r_l} \end{vmatrix} \tag{2}$$

and

$$\varkappa_i = 1 - \sum_{l=1}^n \frac{\frac{1}{p_l} - \frac{1}{p_i}}{r_l} \quad (i = 1, \ldots, n). \tag{3}$$

Let $f(x_1, \cdots, x_n)$, defined on the n-dimensional space R_n, belong to the class $H_{p_1,\cdots,p_n}^{(r_1,\cdots,r_n)}(M)$. Then for any fixed $(x_{m+1}, \cdots, x_n)$ the function f, as a function of $x_1, \cdots, x_m$, belongs to the class $H_q^{(\rho_1,\cdots,\rho_m)}(\overline{M})$. Also

$$\|f\|_{L_q^{(m)}} + \overline{M} < c\left(\min_{1\leqslant i\leqslant n} \|f\|_{L_{p_i}^{(n)}} + M\right),$$

where the constant c is independent of the adjacent factor.

Note. If we set $p = p_i\,(i = 1, \cdots, n)$, then we obtain Theorem 12 of our article [11].

Proof. Let $f \in H_{p_1, \cdots, p_n}^{(r_1, \cdots, r_n)}(M)$. Then by Theorem 2.2 we may associate with f a system of functions $g_{\nu_1, \cdots, \nu_n}$ $(1 \leq \nu_i \leq \infty;\ i = 1, \cdots, n)$, for which the inequalities (2) of §2.2. are valid, as well as similar inequalities obtained by rearranging $x_1, \cdots, x_n$. We will restrict our attention to the case where ν_k depend on an integral parameter s and are defined by the equation

$$\nu_k = \nu_k(s) = 2^{\frac{s}{\rho_k}} \quad (k = 1, \ldots, n,\ s = 0,\ 1, \ldots \text{ and } s = \infty), \tag{4}$$

where the ρ_k are given positive numbers whose precise values will be defined below.

Set

$$\begin{aligned} Q_0 &= g_{\nu_1(0), \ldots, \nu_n(0)}, \\ Q_s &= g_{\nu_1(s), \ldots, \nu_n(s)} - g_{\nu_1(s-1), \ldots, \nu_n(s-1)} \quad (s = 1,\ 2, \ldots). \end{aligned} \tag{5}$$

Obviously

$$Q_s = \sum_{i=1}^{n} Q_s^{(i)} \quad (s = 1,\ 2, \ldots),$$

where

$$Q_s^{(i)} = g_{\nu_1(s), \ldots, \nu_i(s), \nu_{i+1}(s-1), \ldots, \nu_n(s-1)} - g_{\nu_1(s), \ldots, \nu_{i-1}(s), \nu_i(s-1), \ldots, \nu_n(s-1)}. \tag{6}$$

Let q satisfy $p_i \leq q \leq \infty$ $(i = 1, \cdots, n)$, and let m be a natural number such that $1 \leq m \leq n$.

We have

$$\|Q_s\|_{L_q^{(m)}} \leqslant \sum_{i=1}^{n} \|Q_s^{(i)}\|_{L_q^{(m)}} \quad (s = 1,\ 2, \ldots),$$

where $||\psi||_{L_q^{(m)}}$ denotes the norm of ψ computed with respect to the variables $x_1, \cdots, x_m$ for fixed $x_{m+1}, \cdots, x_n$. We apply inequalities (1) and (2) of §1.2 to the ith term of this sum, estimating the norm of the entire function of finite order in the metric of $L_q^{(m)}$, computed in m-dimensional space, using its norm in the metric of $L_{p_i}^{(n)}$, computed in n-dimensional space. Then, noting that $Q_s^{(i)}$ has order not greater than $\nu_i(s)$ with respect to x_i, respectively $(i = 1, \cdots, n)$, we obtain

$$\|Q_s\|_{L_q^{(m)}} \leqslant 2^n \sum_{i=1}^{n} \Big(\prod_{l=m+1}^{n} \nu_l\Big)^{\frac{1}{q}} \|Q_s^{(i)}\|_{L_q^{(n)}}$$

$$\leqslant 4^n \sum_{i=1}^{n} \Big(\prod_{l=m+1}^{n} \nu_l\Big)^{\frac{1}{q}} \Big(\prod_{l=1}^{n} \nu_l\Big)^{\frac{1}{p_i}-\frac{1}{q}} \|Q_s^{(i)}\|_{L_{p_i}^{(i)}} \tag{7}$$

We note that if $\nu_1, \cdots, \nu_n$ and ν_n' are arbitrary finite numbers, where $\nu_n < \nu_n'$, then by the last inequality of (2) of §2.2 we have

$$\|g_{\nu_1,\ldots,\nu_n} - g_{\nu_1,\ldots,\nu_{n-1},\nu_n'}\|_{L_{p_n}^{(n)}} \leqslant \|g_{\nu_1,\ldots,\nu_n} - g_{\nu_1,\ldots,\nu_{n-1},\infty}\|_{L_{p_n}^{(n)}}$$

$$+\|g_{\nu_1,\ldots,\nu_{n-1},\infty} - g_{\nu_1,\ldots,\nu_{n-1},\nu_n'}\|_{L_{p_n}^{(n)}} \leqslant \frac{2cM}{\nu_n^{r_n}} \quad (1 \leqslant \nu_l, \quad l = 1, 2, \ldots, n).$$

Analogously, rearranging the variables, we obtain the more general inequality

$$\|g_{\nu_1,\ldots,\nu_n} - g_{\nu_1,\ldots,\nu_{i-1},\nu_i',\nu_{i+1},\ldots,\nu_n}\|_{L_{p_i}^{(n)}} \leqslant \frac{2cM}{\nu_i^{r_i}}$$

$$(1 \leqslant \nu_l < \infty, \quad \nu_i < \nu', \quad l = 1, \ldots, n).$$

Thus, taking (4) and (6) into account, we will have

$$\|Q_s^{(i)}\|_{L_{p_i}^{(n)}} \leqslant \frac{c_1 M}{2^{\frac{s r_i}{\rho_i}}} \quad (i = 1, \ldots, n, \quad s = 1, 2, \ldots)$$

and consequently (7) implies the following inequality:

$$\|Q_s\|_{L_q^{(m)}} \leqslant c_2 M \sum_{i=1}^{n} 2^{-s\left[\frac{r_i}{\rho_i} - \left(\frac{1}{p_i} - \frac{1}{q}\right)\sum_{l=1}^{n}\frac{1}{\rho_l} - \frac{1}{q}\sum_{m+1}^{n}\frac{1}{\rho_l}\right]},$$

$$(s = 1, 2, \ldots). \tag{8}$$

It is natural to select the constants ρ_l so that the terms on the right-hand side of (8) have the same order. Calculations show that this will happen when

$$\rho_i = \lambda r_i \frac{\varkappa}{\varkappa_i} \quad (i = 1, \ldots, n), \tag{9}$$

where λ is an arbitrary number which we may take equal to 1, and κ and κ_i are the numbers defined by equations (2) and (3) of §2.4.

For ρ_i defined by (9) the bracketed expressions in (8) all reduce to unity, and thus

$$\| Q_s \|_{L_q^{(m)}} \leqslant nc_2 M 2^{-s} = c_3 M 2^{-s} \quad (s = 1, 2, \ldots). \tag{10}$$

Thus these arguments will be correct if the following inequalities are simultaneously satisfied for numbers p_l, r_l and q satisfying $1 \le p_l \le q \le \infty$ and $r_l > 0$:

$$\rho_i = \frac{r_i \varkappa}{\varkappa_i} > 0 \quad (i = 1, \ldots, n).$$

We introduce the series

$$\varphi(x_1, \ldots, x_n) = \varphi = \sum_{s=0}^{\infty} Q_s = g_{\nu_1(0), \ldots, \nu_n(0)}$$

$$+ \sum_{s=1}^{\infty} [g_{\nu_1(s), \ldots, \nu_n(s)} - g_{\nu_1(s-1), \ldots, \nu_n(s-1)}] = \lim_{s \to \infty} g_{\nu_1(s), \ldots, \nu_n(s)}, \tag{11}$$

which we write only formally at this point, without being concerned about convergence.

By (2) of §2.2, where we must assume $\nu_i = \nu_i(s)$ $(i = 1, \cdots, n)$, we may select a subsequence of natural numbers $(s_1, s_2, \cdots)$ such that for almost all $(x_1, \cdots, x_n)$, in the sense of n-dimensional measure, we have

$$\lim_{k \to \infty} g_{\nu_1(s_k), \ldots, \nu_n(s_k)} = f. \tag{12}$$

Thus equation (12) is satisfied on some set $E_1 \subset R_n$ such that the n-dimensional measure of $R_n - E$ is zero.

On the other hand, by (10), the series (11) (for any fixed $x_{m+1}, \cdots, x_n$) converges in the metric of $L_q^{(m)}$ to a function ϕ and thus (in the sense of n-dimensional measure) for almost every point $(x_1, \cdots, x_n)$ we can extract a subsequence $\{s_k'\}$ from $\{s_k\}$ such that at this point

$$\lim_{k \to \infty} g_{\nu_1(s_k'), \ldots, \nu_n(s_k')} = \varphi. \tag{13}$$

Let $E_2 \subset R_n$ be the set on which (13) holds. Obviously on the intersection $E = E_1 E_2$ of the sets E_1 and E_2

$$f = \varphi = \sum_{s=0}^{\infty} Q_s, \tag{14}$$

i.e. almost everywhere in the sense of n-dimensional measure.

By (10), (11) and (14),

$$\| f - g_{\nu_1(s), \ldots, \nu_n(s)} \|_{L_q^{(m)}} \leqslant \sum_{s+1}^{\infty} \| Q_s \|_{L_q^{(m)}} < \frac{c_4 M}{2^s} = \frac{c_4 M}{(2^{\frac{s}{\rho_i}})^{\rho_i}}$$

$$(i = 1, \ldots, m),$$

for each set of values $x_{m+1}, \cdots, x_n$. Since $g_{\nu_1(s), \cdots, \nu_n(s)}$ is an entire function of order $2^{s/\rho_i}$ with respect to x_i for fixed $x_{m+1}, \cdots, x_n$, by Lemma 2.3, the function f belongs to $H_{qx_i}^{(\rho_i)}(\overline{M})$, where

$$\overline{M} < c_5(\|f\|_{L_q^{(m)}} + M) \tag{15}$$

and c_5 is a constant independent of the adjacent factor.

It remains to estimate $\|f\|_{L_q^{(m)}}$.

From (14), by (10),

$$\|f\|_{L_q^{(m)}} \leqslant \|Q_0\|_{L_q^{(m)}} + \sum_1^{\infty} \|Q_s\|_{L_q^{(m)}} < \|Q_0\|_{L_q^{(m)}} + c_5 M \tag{16}$$

Also, by (1) and (2) of §1.2, taking into account the fact that Q_0 has order 1 with respect to each variable $x_i (i = 1, \cdots, n)$, we deduce

$$\|Q_0\|_{L_q^{(m)}} \leqslant 2^n \|Q_0\|_{L_{p_i}^{(n)}} \quad (i = 1, 2, \ldots, n). \tag{17}$$

But, by (3) of §2.1,

$$\|Q_0\|_{L_{p_i}^{(n)}} = \|g_{\nu_1(0), \ldots, \nu_n(0)}\|_{L_{p_i}^{(n)}} \leqslant c \|f\|_{L_{p_i}^{(n)}}. \tag{18}$$

Therefore (16), (17) and (18) imply that

$$\|f\|_{L_q^{(m)}} \leqslant c_6 (\min_{1 \leqslant i \leqslant n} \|f\|_{L_{p_i}^{(n)}} + M). \tag{19}$$

Also, (15) implies that

$$\overline{M} < c_7 (\min_{1 \leqslant i \leqslant n} \|f\|_{L_{p_i}^{(n)}} + M). \tag{20}$$

We have thus proved that, for $\rho_i > 0 (i = 1, \cdots, n)$, the function f, as a function of $x_1, \cdots, x_m$, belongs to the class $H_q^{(\rho_1, \cdots, \rho_m)}(\overline{M})$ and that inequalities (19) and (20) are satisfied, where the constants c_6 and c_7 are independent of the adjacent factors.

2.5. Under the stipulated hypotheses Theorem 2.4 asserts, in particular, that if $f \in H_{p_1, \cdots, p_n}^{(r_1, \cdots, r_n)}$ and $1 \leq m \leq n$, then $f \in H_q^{(\rho_1, \cdots, \rho_m)}$, where the numbers $\rho_i (i = 1, \cdots, m)$ are defined by (1) of §2.4.

Now suppose that we have a pair of numbers m_1 and q_1, where m_1 is a natural number such that $1 \leq m_1 \leq m \leq n$, $q \leq q_1 \leq \infty$ and

$$\mu = 1 - \left(\frac{1}{q} - \frac{1}{q_1}\right) \sum_{k=1}^{m_1} \frac{1}{\rho_k} - \frac{1}{q} \sum_{m_1+1}^{m} \frac{1}{\rho_k} > 0.$$

Then by Theorem 12 of our article [11], the fact that $f \in H_q^{(\rho_1, \cdots, \rho_m)}$ also implies that $f \in H_{q_1}^{(\rho'_1, \cdots, \rho'_{m_1})}$, where

$$\rho'_i = \mu\rho_i, \quad i = 1, \ldots, m_1.$$

Calculations show that if we pass from the numbers n, r_i, $p_i (i = 1, \cdots, n)$ to the numbers m'_1, $\rho'_1, \cdots, \rho'_{m_1}$, q_1, then we again arrive at (1) of §2.4.

Thus the property asserted in Theorem 2.4 has a transitive character, as in the case of our previous theorem ([11], Theorem 12).

2.6. We will prove that under the conditions

$$1 - \sum_1^n \frac{1}{p_l r_l} > 0, \quad 1 < p_i \quad (i = 1, \ldots, n). \tag{1}$$

Theorem 2.4 cannot be improved (in terms that apply here). Namely, for any number p' such that $1 \leq p_i \leq p' \leq \infty$ and any natural number m such that $1 \leq m \leq n$, we can find a function f in the class $H_{p_1, \cdots, p_n}^{(r_1, \cdots, r_n)}$, but not in the class $H_q^{(\rho_1, \cdots, \rho_{i-1}, \rho_i + \epsilon, \rho_{i+1}, \cdots, \rho_m)}$ (for any M), where the numbers $\rho_l (l = 1, \cdots, m)$ are defined by (2) of §2.4.

It is easy to see that if we can find an example of the required function f for $q = \infty$ and $m = 1$ then, by the transitive property of Theorem 2.4 (see §2.5), this function can also serve as the required example for other values of m, where $1 \leq m \leq n$ and for p', where $p_i \leq p' \leq \infty$.

Thus, let $r_i > 0$ and p_i satisfy $1 < p_i \leq \infty$. Moreover, suppose (1) holds.

We will give the function f in the form of a series

$$f = f(x_1, \ldots, x_n) = \sum_{\nu=0}^{\infty} \varphi_\nu, \tag{2}$$

where

$$\varphi_\nu = \varphi_\nu(x_1, \ldots, x_n) = \frac{1}{2^\nu} \prod_{i=1}^{n} \frac{\sin 2^{\nu\lambda_i} x_i}{x_i} \quad (\nu = 1, 2, \ldots), \tag{3}$$

and the numbers λ_i are defined by

$$\lambda_i = \frac{1}{r_i}\left(1 - \frac{1}{q_i} \frac{\sum_{l=1}^{n} \frac{1}{r_l}}{1 + \sum_{l=1}^{n} \frac{1}{q_l r_l}}\right) = \frac{1 - \sum_1^n \frac{1}{r_l p_l} + \frac{1}{p_i} \sum_{l=1}^{n} \frac{1}{r_l}}{r_i \left(1 - \sum_{l=1}^{n} \frac{1}{r_l p_l} + \sum_1^n \frac{1}{r_l}\right)},$$

$$\frac{1}{p_l} + \frac{1}{q_l} = 1 \quad (i, l = 1, \ldots, n). \tag{4}$$

By (1) the numbers λ_i are positive.

By the equations

$$\left(\int_{-\infty}^{\infty}\left|\frac{\sin nx}{x}\right|^p dx\right)^{\frac{1}{p}} = n^{\frac{1}{q}}\left(\int_{-\infty}^{\infty}\left|\frac{\sin x}{x}\right|^p dx\right)^{\frac{1}{p}} = n^{\frac{1}{q}} A$$

$$\left(\frac{1}{p}+\frac{1}{q}=1\right),$$

we have

$$\left\| f - \sum_1^{\mu-1}\varphi_\nu \right\|_{L_{p_i}^{(n)}} \leqslant \sum_{\mu}^{\infty} \|\varphi_\nu\|_{L_{p_i}^{(n)}} = \sum_{\mu}^{\infty} \frac{A}{2^{\nu\left(1-\sum_{l=1}^{n}\frac{\lambda_l}{q_l}\right)}}$$

$$= \sum_{\mu}^{\infty}\frac{A}{2^{\nu\lambda_i r_i}} < \frac{c}{2^{\mu\lambda_i r_i}} = \frac{c}{(2^{\mu\lambda_i})^{r_i}} \qquad (i=1,\ldots,n),$$

and since the function $\Sigma_1^{\mu-1}\phi_\nu$ is an entire function of order $2^{\nu\lambda_i}$ with respect to x_i, Lemma 2.3 implies

$$f \in H_{p_1,\ldots,p_n}^{(r_1,\ldots,r_n)}.$$

For $p' = \infty$ and $m = 1$ the hypotheses of Theorem 2.4 regarding ρ_i and r_i are satisfied, since in this case

$$\varkappa_1 = 1 - \sum_{l=1}^{n}\frac{1}{p_l r_l} + \frac{1}{p_1}\sum_{l=1}^{n}\frac{1}{r_l} > 1 - \sum_{l=1}^{n}\frac{1}{p_l r_l} = \varkappa > 0$$

and consequently

$$\rho_1 = \frac{r_1\varkappa}{\varkappa_1} > 0.$$

Thus by Theorem 2.4,

$$\psi(x) = f(x,0,\ldots,0) = \sum_{\nu=0}^{\infty}\frac{1}{2^{\nu\left(1-\sum_1^n \lambda_i\right)}}\frac{1}{x}\int_0^x \cos 2^{\nu\lambda_1}u\,du$$

$$= \frac{1}{x}\int_2^x \sum_{\nu=0}^{\infty}\frac{\cos 2^\nu u}{2^{\rho_1\nu}}\,du = F_{\rho_1}(x) \in H_{\infty x_1}^{(\rho_1)}.$$

But in [11] we prove that the function $F_{\rho_i}(x)$ belongs to none of the classes $H_{\infty x}^{(\rho_1+\epsilon)}$ for $\epsilon > 0$ and the function f defined by (2) can thus serve as the example, corroborating the assertion made at the beginning of §2.6.

2.61. The example of §2.6 covers only those cases provided for in

Theorem 2.4 for which (1) of §2.6 is also satisfied. A more general example covering all cases can also be constructed in the form of a series similar to (2) based on the corresponding results of T. I. Amanov [1], who in a comprehensive way supplemented my previously cited result ([11], Theorem 12) with the missing examples.

§3

3.1. Suppose $H_{p_1, \cdots, p_n}^{(r_1, \cdots, r_n)}$ denotes that set of functions f that belong to $H_{p_1, \cdots, p_n}^{(r_1, \cdots, r_n)}(M)$ for some M, and let M_f denote the smallest constant M for which this occurs.

For each function $f \in H_{p_1, \cdots, p_n}^{(r_1, \cdots, r_n)}$ we introduce the norm[1)]

$$\|f\|_{L_{p_1, \ldots, p_n}}^{(r_1, \ldots, r_n)} = \max_{1 \leqslant i \leqslant n} \|f\|_{L_{p_i}^{(n)}} + M_f.$$

The set $H_{p_1, \cdots, p_n}^{(r_1, \cdots, r_n)}$ with the norm we have introduced forms a complete normed linear space, i.e. a Banach space. The proof of this assertion is analogous to the proof for $H_p^{(r_1, \cdots, r_n)}$ in our article [15, §4].

3.2. **Compactness Theorem.** *Let $\{f_m\}$ be a sequence of functions satisfying*

$$f_m \in H_{p_1, \ldots, p_n}^{(r_1, \ldots, r_n)}, \qquad \|f_m\|_{L_{p_1, \ldots, p_n}}^{(r_1, \ldots, r_n)} \leqslant K \quad (m = 1, 2, \ldots), \tag{1}$$

where K is a given constant, and let the conditions of Theorem 2.4 be satisfied.

Then we can select a subsequence $\{f_{m_\nu}\}$ and a function f satisfying (1) *such that for all numbers ρ_i' for which $0 < \rho_i' < \rho_i$ $(i = 1, \cdots, n)$ and for every bounded domain $G \subset R_n$,*

$$\lim \|f_{m_\nu} - f\|_{L_q, (K_m G)}^{(\rho_1', \cdots, \rho_n')} = 0. \tag{2}$$

The proof of this theorem depends on our theorem that the classes $H_p^{(r_1, \cdots, r_n)}$ are compact [14].

§4

4.1. In order to apply the above theorems to functions defined on any domain G in R_n, we can try to continue the given function f from G to the whole space R_n in such a way that the continuation belongs to one of the classes $H_{p_1, \cdots, p_n}^{(r_1, \cdots, r_n)}$.

[1)]Similarly we can introduce the space $H_{p_1, \cdots, p_n}^{(r_1, \cdots, r_n)}(G)$ of functions defined on a domain $G \subset R_n$, which will be a normed linear space and for certain domains will also be complete.

Here we only note the following case. Together with the space R_n of points $(x_1, \cdots, x_n)$ we define the subspace R' of points $(x_1, \cdots, x_m)$ and the subspace R'' of points $(x_{m+1}, \cdots, x_n)$. Let $G' \subset R'$ and $G'' \subset R''$ be domains bounded by surfaces that are continuously differentiable ρ_1 and ρ_2 times, respectively, and $G = G' \times G'' \subset R$ be the topological product of G' and G''. We define a function $f(x_1, \cdots, x_n)$ on G with the property that it together with its generalized partial derivatives of the form

$$\frac{\partial^{\rho'} f}{\partial x_1^{\alpha_1} \ldots \partial x_m^{\alpha_m}} \qquad \left(\sum_{k=1}^{m} \alpha_K = \rho'\right)$$

is p'-integrable on G, and, moreover, it together with its generalized partial derivatives of the form

$$\frac{\partial^{\rho''} f}{\partial x_m^{\alpha_{m+1}} \ldots \partial x_n^{\alpha_n}} \qquad \left(\sum_{m+1}^{n} \alpha_K = \rho''\right)$$

is p''-integrable on G $(1 \leq p'' \leq \infty)$.

Let the functions f defined above form the class $W^{(\rho', \rho'')}_{p'(G'), p''(G'')}$. Set

$$\|f\|_{W^{(\rho', \rho'')}_{p'(G'), p''(G'')}} = \max \{\|f\|_{L_{p'}(G)}, \|f\|_{L_{p''}(G)}\} + \sum \|f^{(\rho')}\|_{L_{p'}(G)} + \sum \|f^{(\rho'')}\|_{L_{p''}(G'')},$$

where the sums are extended over all partial derivatives of f of order ρ' with respect to the variables $x_1, \cdots, x_m$ and over all partial derivatives of order ρ'' of f with respect to the variables $x_{m+1}, \cdots, x_n$.

A function f in the class $W^{(\rho', \rho'')}_{p'(G'), p''(G'')}$, can be continued from G to R_n in such a way that the continuations $\bar{f}$ will belong to the class $W^{(\rho', \rho'')}_{p'(R_m), p''(R_{n-m})}$, will equal 0 outside some n-dimensional ball and will satisfy

$$\|\bar{f}\|_{W^{(\rho', \rho'')}_{p'(R_m), p''(R_{n-m})}} < c \|f\|_{W^{(\rho', \rho'')}_{p'(G'), p''(G'')}},$$

where c is a constant independent of the adjacent factor. Here if f is bounded on G by a constant K the continuation will be bounded by cK, where c is independent of K.

We will not carry out the proof of this assertion since it is only slightly more complicated than the proof of the continuation theorem for the classes $W_p^{(\rho)}(G)$ [13, 14, 15, 2].

It is not difficult to see that

$$W^{(\rho', \rho'')}_{p'(R_m), p''(R_{n-m})} \subset H^{(r_1, \ldots, r_n)}_{p_1, \ldots, p_n},$$

where

$$p' = p_1 = \cdots = p_m, \quad p'' = p_{m+1} = \cdots = p_n,$$

$$\rho' = r_1 = \cdots = r_m, \quad \rho'' = r_{m+1} = \cdots = r_n,$$

and also

$$\| \bar{f} \|_{L_{p_1, \ldots, p_n}}^{(r_1, \ldots, r_n)} < c_1 \| \bar{f} \|_{W_{p'(R_m), p''(R_{n-m})}}^{(\rho', \rho'')},$$

where c_1 is a constant independent of the adjacent factor.

Of course similar results hold for the case when the domain G is the topological product of not two but several domains.

4.2. We consider still another elementary example. Let the set $\mathfrak{E}$ consist of the functions f defined and bounded on the parallelepiped $G = \{a_i \leq x_i \leq b_i, i = 1, \cdots, n\}$ such that inequality (1) of §1.1 holds. Since G is the topological product of the segments $[a_i, b_i]$, by the above result we can assert that the functions $f \in \mathfrak{E}$ can be continued to R_n in such a way that the extensions $\bar{f}$ will satisfy the inequality

$$\| \bar{f} \|_{L_{p_i}^{(n)}} + \int_{R_n} \sum_{i=1}^{n} \left| \frac{\partial^{r_i} \bar{f}}{\partial x_i^{r_i}} \right| dR_n < M_1, \tag{1}$$

where M_1 is a constant. It is easy to see that the functions $\bar{f}$ will also belong to the space $H_{p_1, \cdots, p_n}^{(r_1, \cdots, r_n)}$ ([11], Lemma 3) and that their norms in the sense of this space will be bounded:

$$\| \bar{f} \|_{L_{p_1, \ldots, p_n}}^{(r_1, \ldots, r_n)} \leqslant M_r.$$

Now suppose that

$$1 - \sum_{1}^{n} \frac{1}{p_i r_i} > 0.$$

We apply Theorem 2.4, where $q = \infty$. We have

$$\varkappa = 1 - \sum_{1}^{n} \frac{1}{p_l r_l} > 0,$$

$$\varkappa_i = \varkappa + \frac{1}{p_i} \sum_{l=1}^{n} \frac{1}{r_l} > 0 \quad (i = 1, \ldots, n).$$

Thus

$$\rho_i = \frac{r_i \varkappa}{\varkappa_i} > 0 \quad (i = 1, \ldots, n),$$

and the hypotheses of Theorem 2.4 are satisfied. We can therefore assert that the functions $\bar{f}$, the continuations of the functions $f \in \mathfrak{S}$, belong to the class $H_{\infty}^{(\rho_1, \cdots, \rho_n)}$, and that

$$\|\bar{f}\|_{L_{\infty}^{(n)}}^{(\rho_1, \ldots, \rho_n)} < c \|\bar{f}\|_{L_{p_1, \ldots, p_n}^{(n)}}^{(r_1, \ldots, r_n)} \leqslant M_2.$$

This means, in particular, that $\mathfrak{S}$ is a bounded and equicontinuous set.

4.3. Finally we note that we may consider the classes $H_{* p_1, \cdots, p_n}^{(r_1, \cdots, r_n)}$ of functions that are periodic with period 2π in each variable x_i ($i = 1, \cdots, n$), where the norms of the functions are calculated with respect to the period. We can obtain a complete theory for these classes analogous to the one presented by us for the classes $H_{p_1, \cdots, p_n}^{(r_1, \cdots, r_n)}$. The role of the entire functions of finite order will be played by the trigonometric polynomials $T_{\nu_1, \cdots, \nu_n}$ of order ν_i with respect to x_i, respectively ($i = 1, \cdots, n$). In particular, we must take $G = \{0 \leq x_i \leq 2\pi;\ i = 1, \cdots, n\}$ in (2) of the compactness theorem.

BIBLIOGRAPHY

[1] T. I. Amanov, *Boundary functions of the classes* $H_p^{(r_1, \cdots, r_n)}$ *and* $H_{p*}^{(r_1, \cdots r_n)}$, Izv. Akad. Nauk SSSR Ser Mat. 19 (1955), 17–32. (Russian) MR 16, 806.

[2] V. M. Babič, *On the extension of functions*, Uspehi Mat. Nauk 8 (1953), no. 2 (54), 111–113. (Russian) MR 15, 110.

[3] N. K. Bari, *Generalization of inequalities of S. N. Bernšteĭn and A. A. Markov*, Izv. Akad. Nauk SSSR Ser. Mat. 18 (1954), 159–176. (Russian) MR 15, 788.

[4] S. N. Bernšteĭn, *On properties of homogeneous functional classes*, Dokl. Akad. Nauk SSSR 57 (1947), 111–114. (Russian) MR 9, 235.

[5] ———, *On the best approximation of functions of several variables by means of polynomials of trigonometric sums*, Trudy Mat. Inst. Steklov. 38 (1951), 24–29. (Russian) MR 13, 842.

[6] D. Jackson, *Theory of approximation*, Amer. Math. Soc. Colloq. Publ., Vol. 11, Amer. Math. Soc., Providence, R. I., 1930.

[7] E. Gagliardo, *Un criterio di eguale continuità per funzioni di due variabili*, Rend. Sem. Mat. Univ. Padova 26 (1956), 148–167. MR 18, 878.

[8] I. I. Ibragimov, *Extremal problems in a class of entire functions of exponential type*, Uspehi Mat. Nauk 12 (1957), no. 3 (75), 323–328. (Russian) MR 19, 737.

[9] W. Kondrachov, *Sur certaines propriétés des fonctions dans l'espace* L_p^{ν}, C. R. (Dokl.) Acad. Sci. URSS 48 (1945), 35–38. MR 8, 32.

[10] S. M. Nikol'skiĭ, *Some inequalities for entire functions of finite degree of several variables and their application,* Dokl. Akad. Nauk SSSR 76 (1951), 785–788. (Russian) MR 12, 603.

[11] ———, *Inequalities for entire functions of exponential type and their application to the theory of differentiable functions of several variables,* Trudy Mat. Inst. Steklov. 38 (1951), 244–278; English transl., Amer. Math. Soc. Transl. (2) 80 (1969), 1–38. MR 14, 32.

[12] ———, *Properties of certain classes of functions of several variables on differentiable manifolds,* Mat. Sb. 33 (75) (1953), 261–326; English transl., Amer. Math. Soc. Transl. (2) 80 (1969), 39–118. MR 16, 453.

[13] ———, *On the solution of the polyharmonic equation by a variational method,* Dokl. Akad. Nauk SSSR 88 (1953), 409–411. (Russian) MR 15, 425.

[14] ———, *Compactness of the classes* $H_p^{(r_1,\dots,r_n)}$ *of functions of several variables,* Izv. Akad. Nauk SSSR Ser. Mat. 20 (1956), 611–622. (Russian) MR 18, 321.

[15] ———, *Extension of functions of several variables preserving differential properties,* Mat. Sb. 40 (82) (1956), 243–268; English transl., Amer. Math. Soc. Transl. (2) 83 (1969), 159–188. MR 19, 398.

[16] S. L. Sobolev, *On a theorem of functional analysis,* Mat. Sb. 4 (46) (1938), 471–497; English transl., Amer. Math. Soc. Transl. (2) 34 (1963), 39–68.

[17] ———, *Applications of functional analysis in mathematical physics,* Izdat. Leningrad. Gos. Univ., Leningrad, 1950; English transl., Transl. Math. Monographs, vol. 7, Amer. Math. Soc, Providence, R. I., 1963. MR 14, 565; 29 #2624.

Translated by:
A. Shuchat

Amer. Math. Soc. Transl.
(2) Vol. 90, 1970

IMBEDDING THEOREMS FOR GENERALIZED SOBOLEV CLASSES W_p^r

S. V. USPENSKIĬ *

In solving Dirichlet's problem for differential equations of elliptic type by a variational method, the question arises of what necessary and sufficient conditions must be imposed on the boundary function in order that an admissible function of the variational problem exist. It turns out that these conditions can be expressed in the form of specific integral equations, and the classes of functions which satisfy them are a natural generalization to nonintegral indices of the well-known Sobolev classes W_p^r (see [1-3]). These classes became the subject of numerous investigations by various authors, most attention being focused on so-called imbedding theorems or on determining the differential properties of the class $W_p^r(G^n)$ in the $L_{p_1}(G^m)$-metric ($p_1 \geq p$, $m \leq n$, and G^m is an m-dimensional manifold belonging to G^n). Direct imbedding theorems for the classes W_p^r which are invertible (i.e. expressed in terms which are necessary and sufficient) were first obtained for $r = 1$, $p = 2$ by Aronszajn [4], Slobodeckiĭ and Babič [5]. In [2,6] Slobodeckiĭ generalized this result for $p = 2$ to all indices $r > 0$. However, as Besov showed [9], these results do not carry over to the case $p \neq 2$. The case $p \neq 2$ required the introduction of new ideas and was examined in articles by Gagliardo [7], Slobodeckiĭ [8], Besov [9], the author [10-11], and Solonnikov and Il'in [26].

This paper presents a completed theory of imbeddings for the generalized Sobolev classes W_p^r. The proofs use a new method, which differs from those of Sobolev, Nikol'skiĭ and Gagliardo. Most of the results below were published without proof in our notes [10-11]. The paper consists of five sections. The first three are preliminary. The fundamental results of the paper are contained in §§4 and 5. The paper uses both the notation and the results of the author's article [12].

§ 1. Basic definitions

Let R^n be the n-dimensional space of points $(x_1, \cdots, x_n)$, and let R^m be the m-dimensional space of points $(x_1, \cdots, x_m, x_{m+1}^0, \cdots, x_n^0)$,

*) Translation of Sibirsk. Mat. Ž. 3 (1962), 418–445.

where $x^0_{m+1}, \cdots, x^0_n$ are fixed. Further, let $\overline{R^n_\rho}$ be the half-space $\{x_n \geq \rho, x(x_1, \cdots, x_{n-1}) \in R^{n-1}\}$, and let R^m_ρ be the hyperplane $x_{m+1} = \cdots = x_n = \rho$.

Following S. M. Nikol'skiĭ, we shall say that the function f has a generalized derivative $\partial f/\partial x_i$ $(i = 1, 2, \cdots, n)$ in the region $G \subset R^n$ if f can be altered on a set of measure zero in such a way that it is absolutely continuous in x_i for almost all admissible $(x_1, \cdots, x_{i-1}, x_{i+1}, \cdots, x_n)$. The derivative $\partial^l f/\partial x_1^{\alpha_1} \cdots \partial x_n^{\alpha_n}$ $(l = \alpha_1 + \cdots + \alpha_n)$ is defined by induction. Since we shall always assume that all the generalized partial derivatives of order $k < l$ exist and are integrable on G if the generalized derivative $\partial^l f/\partial x_1^{\alpha_1} \cdots \partial x_n^{\alpha_n}$ exists and is integrable on G, it follows that this definition of the generalized derivative coincides with Sobolev's (see [1,20,21]). We introduce the classes $L^i_{p,\alpha}(\overline{R}^n_0)$ and $W^l_{p,\alpha}(\overline{R}^n_0)$ where l is an integer, $1 < p < \infty$, $\alpha > -1$. The function f belongs to the class $L^l_{p,\alpha}$ if

1) f has all generalized derivatives through the lth order on $\overline{R}^n_0$;

2)

$$\|f\|^p_{L^l_{p,\alpha}(\overline{R}^n_0)} = \int_{\overline{R}^n_0} x_n^\alpha \sum_{\alpha_1+\cdots+\alpha_n=l} \left| \frac{\partial^l f}{\partial x_1^{\alpha_1} \cdots \partial x_n^{\alpha_n}} \right|^p d\overline{R}^n_0 < \infty. \tag{1}$$

The function f belongs to $W^l_{p,\alpha}(\overline{R}^n_0)$ if $f \in L^l_{p,\alpha}$ and

$$\|f\|^p_{L_p(\overline{R}^n_0)} = \int_{\overline{R}^n_0} |f|^p d\overline{R}^n_0 < \infty.$$

We set

$$\|f\|_{W^l_{p,\alpha}(\overline{R}^n_0)} = \|f\|_{L_p(\overline{R}^n_0)} + \|f\|_{L^l_{p,\alpha}(\overline{R}^n_0)}.$$

We define the generalized Sobolev classes $W^r_p(R^n)$ $(1 < p < \infty$, $r = \overline{r} + \alpha$, $\overline{r}$ is integral, $0 \leq \alpha < 1)$. Let r be integral. We shall say that f belongs to the class $W^r_p(R^n)$ if f has all generalized Sobolev derivatives through the rth order, inclusive, and

$$\|f\|^p_{W^r_p(R^n)} = \int_{R^n} \sum_{\alpha_1+\cdots+\alpha_n=r} \left| \frac{\partial^r f}{\partial x_1^{\alpha_1} \cdots \partial x_n^{\alpha_n}} \right|^p dR^n + \int_{R^n} |f|^p dR^n < \infty.$$

When r is not an integer $(0 < \alpha < 1)$, we shall say that $f \in W_p^r(R^n)$, if

1) f belongs to the class $W_p^{\bar{r}}(R^n)$;

2)

$$\| f^{\bar{r}}_{\alpha_1 \ldots \alpha_n} \|^p_{\overline{W}^\alpha_p(R^n)} = \sum_{j=1}^{n} \int_0^\infty \int_{R^n} \frac{| \Delta_{j,h} f^{\bar{r}}_{\alpha_1 \ldots \alpha_n} |^p}{h^{1+p\alpha}} \, dR^n dh < \infty. \qquad (2)$$

Here $\bar{r} = \alpha_1 + \cdots + \alpha_m$, $f^{\bar{r}}_{\alpha_1 \ldots \alpha_n}$ is the generalized Sobolev derivative, and

$$\Delta_{j,h} f = f(x_1, \ldots, x_j + h, \ldots, x_n) - f(x_1, \ldots, x_j, \ldots, x_n).$$

We set

$$\| f \|_{W_p^r(R^n)} = \| f \|_{W_p^{\bar{r}}(R^n)} + \sum_{\alpha_1 + \ldots + \alpha_n = \bar{r}} \| f^{\bar{r}}_{\alpha_1 \ldots \alpha_n} \|_{\overline{W}^\alpha_p(R^n)}. \qquad (3)$$

As will be shown below, these generalized Sobolev classes $W_p^r(R^n)$ coincide with the classes $W_p^r(R^n)$ introduced by Slobodeckiĭ in [2,3]. We define the classes W_p^r on the half-space R_0^n similarly.

We further define the function classes $B_p^r(R^n)$ $(r = \bar{r} + \alpha$, $\bar{r}$ is integral, $0 < \alpha \leq 1$, $p > 1)$, For r not an integer, we shall say that $f \in B_p^r(R^n)$ if $f \in W_p^r(R^n)$. For r an integer $(\alpha = 1)$, we shall say that $f \in B_p^r$ if $f \in W_p^{r-1}(R^n)$ and

$$\| f^{r-1}_{\alpha_1 \ldots \alpha_n} \|^p_{\overline{B}^1_p} = \sum_{j=1}^{n} \int_0^\infty \int_{R^n} \frac{| \Delta^2_{j,h} f^{r-1}_{\alpha_1 \ldots \alpha_n} |^p}{h^{1+p}} \, dR^n dh < \infty, \qquad (4)$$

where $\Delta^2_{j,h} f = f(x_1, \cdots, x_j - h, \cdots, x_n) - 2f(x_1, \cdots, x_n) + f(x_1, \cdots, x_j + h, \cdots, x_n)$.

For r an integer we set

$$\| f \|_{B_p^r} = \| f \|_{W_p^{r-1}(R^n)} + \sum_{\alpha_1 + \ldots + \alpha_n = r-1} \| f^{r-1}_{\alpha_1 \ldots \alpha_n} \|_{\overline{B}^1_p}.$$ [1)]

We shall say that a function F in $L_p(R^n)$ takes the values ϕ on R^m $(F|_{R^m} = \phi)$ if, possibly after altering F on a set of n-dimensional measure

1) As Besov showed, for r not an integer this definition is equivalent to the definition of the classes $B^r_{p,p}$ in [9]. Theorems 7 and 8 imply that these definitions are also equivalent in the integral case.

zero, we have

$$\lim_{x_i \to x_i^0} \int_{R^m} |F - \varphi|^p dR^m = 0, \; i = m+1, \; \ldots \; n. \tag{5}$$

We shall further say that the function class W_p^l belongs to the class $W_q^r(B_q^r)$ if every function $f \in W_p^l$ satisfies

$$\|f\|_{W_q^r(B_q^r)} \leqslant c\|f\|_{W_p^l},$$

where the constant c does not depend on f.

We state some theorems needed later, whose proofs can be found in the author's article [12].

Theorem 1. *If* $\alpha - kp + 1 > 0$, *then*

$$W_{p,\alpha}^l(\overline{R}_0^n) \subset W_{p,\alpha-kp}^{l-k}(\overline{R}_0^n) \tag{6}$$

and any function $v \in W_{p,\alpha}^l(R_0^n)$ *satisfies*

$$\lim_{\substack{\rho \to 0 \\ (\rho \to \infty)}} \rho^{\alpha - kp + 1} \cdot \int_{R_\rho^{n-1}} \sum_{\alpha_1 + \ldots + \alpha_n = l-k} \left| \frac{\partial^{l-k} v}{\partial x_1^{\alpha_1} \ldots \partial x_n^{\alpha_n}} \right|^p dR_\rho^{n-1} = 0.$$

Corollary. *For* $v \in L_{p,\alpha}^l(\overline{R}_0^n)$ *let*

$$\lim_{\rho \to \infty} \int_{R_\rho^{n-1}} \sum_{\alpha_1 + \ldots + \alpha_n = k} \left| \frac{\partial^k v}{\partial x_1^{\alpha_1} \ldots \partial x_n^{\alpha_n}} \right|^p dR_\rho^{n-1} = 0. \quad k = 0, 1, \ldots, l-1.$$

Then, if $\alpha - kp + 1 > 0$, *it follows that*

$$\|v\|_{L_{p,\alpha-kp}^l(\overline{R}_0^n)} \leqslant c\|v\|_{L_{p,\alpha}^l(\overline{R}_0^n)}, \tag{7}$$

where the constant c *does not depend on* v.

Theorem 2. *If* $\alpha - pl + 1 < 0$, *then*

$$W_{p,\alpha}^l(\overline{R}_0^n) \subset B_p^{l - \frac{\alpha+1}{p}}(R_0^{n-1}). \tag{8}$$

Theorem 3 (converse). *Let* $f \in B_p^r(R_0^{n-1})$. *Then the harmonic function*

$$U = kx_n \int\limits_{R^{n-1}} \frac{f}{[(t-x)^2 + x_n^2]} dt \tag{9}$$

satisfies the inequality

$$\| U \|_{L^l_{p, p(l-r)-1}(\overline{R}^n_0)} \leqslant c \| f \|_{B^r_p(R^{n-1}_0)}, \tag{10}$$

where c *does not depend on* f.

Theorem 4. *If* $0 \leq \alpha \leq pl$, *then*

$$W^l_{p,\alpha}(\overline{R}^n_0) \subset W_p^{l-\frac{\alpha}{p}}(\overline{R}^n_0). \tag{11}$$

§ 2. Properties of harmonic functions in a half-space. Littlewood-Paley theorem

In [13] Littlewood and Paley found some estimates for harmonic functions in a disk ($n = 2$). Below this result is generalized to the half-space $x_n \geq 0$ ($n \geq 2$). We shall say that a harmonic function U defined on the half-space $x_n \geq 0$ satisfies conditions a) and b) if

a) $U|_{R^{n-1}} = f \in L_p(R^{n-1})$;

b) $\int\limits_{R^{n-1}_\rho} |U|^p dR^{n-1}_\rho < M, \ 0 < \rho < \infty.$

Lemma 1. *Conditions* a) *and* b) *determine a unique harmonic function; it has the representation*

$$U = kx_n \int\limits_{R^{n-1}} \frac{f}{[(t-x)^2 + x_n^2]^{\frac{n}{2}}} dt. \tag{12}$$

Here

$$k = \left[\int\limits_{R^{n-1}} \frac{dt}{[t^2+1]^{\frac{n}{2}}} \right]^{-1}, \quad x = x(x_1, \ldots, x_{n-1}), \ t = t(t_1, \ldots, t_{n-1}).$$

Proof (suggested by S. M. Nikol'skiĭ). Let $\rho > 0$.

For all points $x(x_1, \cdots, x_n) \in \overline{R}^n_\rho$ we have

$$|U(x_1, \ldots, x_n)| \leqslant \frac{c}{\rho^n} \int\limits_{\omega_\rho} |U| \, d\omega_\rho \leqslant$$

$$\leqslant c\rho^{-\frac{n}{p}}\left[\int\limits_{\omega_\rho}|U|^p d\omega_\rho\right]^{\frac{1}{p}} \leqslant c\rho^{-\frac{n}{p}+1}\sup_u\left\{\int\limits_{R_u^{n-1}}|U|^p dR_u^{n-1}\right\},$$

where ω_ρ is the n-dimensional ball with center at $(x_1, \cdots, x_n)$ and radius ρ. Then (see, for example, [14]) in $\bar{R}^n_\rho$ the harmonic function U has the form

$$U = k(x_n-\rho)\int\limits_{R^{n-1}}\frac{U(t, \rho)\,dt}{[(t-x)^2+(x_n-\rho)^2]^{\frac{n}{2}}}.$$

We set

$$V_\rho = k(x_n-\rho)\int\limits_{R^{n-1}}\frac{f dt}{[(t-x)^2+(x_n-\rho)^2]^{\frac{n}{2}}}.$$

For $x_n > \rho$ we have

$$\int\limits_{R^{n-1}}|U-V_\rho|^p dR^{n-1} \leqslant \int\limits_{R^{n-1}}|U_\rho - f|^p dR^{n-1} \to 0, \tag{13}$$

$$\text{if}\quad \rho\to 0.$$

To prove the assertion, it suffices to note that

$$V_\rho \to kx_n\int\limits_{R^{n-1}}\frac{f dt}{[(t-x)^2+x_n^2]^{\frac{n}{2}}}\quad\text{as}\quad \rho\to 0.$$

Lemma 2. *Let* $f\in L_p(R^{n-1})$, $p>1$, *and*

$$U = kx_n\int\limits_{R^{n-1}}\frac{f dt}{[(t-x)^2+x_n^2]^{\frac{n}{2}}}.$$

Then

$$\int\limits_{R_\rho^{n-1}}|U^l_{\alpha_1\ldots\alpha_n}|^p dR_\rho^{n-1} = o(\rho^{-pl})\ (\rho\to\infty), \tag{14}$$

$$\int\limits_{R_\rho^{n-1}}|U^{l+1}_{\alpha_1\ldots\alpha_n}|^p dR_\rho^{n-1} = o(\rho^{-p(l+1)})\ (\rho\to 0), \tag{15}$$

where

$$U^l_{\alpha_1\ldots\alpha_n} = \frac{\partial^l U}{\partial x_1^{\alpha_1}\ldots\partial x_n^{\alpha_n}}, \quad \alpha_1+\ldots+\alpha_n = l, \ l = 0, 1, \ldots .$$

Proof. It is easy to see that

$$\left|\frac{\partial^l}{\partial x_1^{\alpha_1}\ldots\partial x_n^{\alpha_n}}\left(\frac{x_n}{[(t-x)^2+x_n^2]^{\frac{n}{2}}}\right)\right| \leqslant c\,\frac{1}{[(t-x)^2+x_n^2]^{\frac{n+l-1}{2}}}, \tag{16}$$

where the constant c depends only on l and n.

We then have

$$\left|\frac{\partial^l U}{\partial x_1^{\alpha_1}\ldots\partial x_n^{\alpha_n}}\right| \leqslant c \int\limits_{R^{n-1}} \frac{|f|\,dt}{[(t-x)^2+x_n^2]^{\frac{n+l-1}{2}}} \leqslant c x_n^{1-l} \int\limits_{R^{n-1}} \frac{|f|\,dt}{[(t-x)^2+x_n^2]^{\frac{n}{2}}}$$

$$\times \int\limits_{R^{n-1}} |U^l|\,dR^{n-1} \leqslant c\cdot x_n^{-p(l-1)} \int\limits_{R^{n-1}} \left[\int\limits_{R^{n-1}} \frac{|f|\,dt}{[(t-x)^2+x_n^2]^{\frac{n}{2}}}\right]^p dR^{n-1}$$

$$\leqslant c x_n^{-p(l-1)} \int\limits_{R^{n-1}} \left[\int\limits_{R^{n-1}} \frac{|f|^p dt}{[(t-x)^2+x_n^2]^{\frac{n}{2}}}\right]\left[\int\limits_{R^{n-1}} \frac{dt}{[(t-x)^2+x_n^2]^{\frac{n}{2}}}\right]^{p-1} dR^{n-1}$$

$$= c x_n^{1-pl} \int\limits_{R^{n-1}}\int\limits_{R^{n-1}} \frac{|f|^p}{[(t-x)^2+x_n^2]^{\frac{n}{2}}}\,dt\,dR^{n-1} = c x_n^{-pl} \int\limits_{R^{n-1}} |f|^p dR^{n-1}. \tag{17}$$

(17) implies compactness of the set of functions

$$\{U\} = \{U(x, \rho)\} \text{ on } R^{n-1} \text{ for } \rho > \rho_0 > 0.$$

Let Δ_N be the region $(-N \le x_i \le N,\ i = 1, \cdots, n-1)$. Then, since $\{U_\rho\}$ is compact, for any $\epsilon > 0$ there is a number N_1, which depends only on ϵ, but not on x_n, such that for $N > N_1$ we have

$$\int\limits_{R^{n-1}-\Delta_N} \left(k x_n \int\limits_{R^{n-1}} \frac{|f|\,dt}{[(t-x)^2+x_n^2]^{\frac{n}{2}}}\right)^p dx < \frac{\epsilon}{3}.$$

On the other hand, since $f \in L_p(R^{n-1})$, there is a number N_2, which depends only on N_1 and ϵ, such that for $N > N_2$

$$\int_{\Delta_N} \left(kx_n \int_{R^{n-1}-\Delta_N} \frac{|f|\,dt}{[(t-x)^2+x_n^2]^{\frac{n}{2}}} \right)^p dx < \frac{\varepsilon}{3}.$$

But this easily implies (14). We prove (15). To be definite we set $n > 2$:

$$\frac{\partial^l U}{\partial x_1^{\alpha_1} \ldots \partial x_n^{\alpha_n}} = \frac{k(-1)^{l-\alpha_n}}{n-2} \int_{R^{n-1}} f \frac{\partial^{l+1}}{\partial t_1^{\alpha_1} \ldots \partial t_{n-1}^{\alpha_{n-1}} \partial x_n^{\alpha_n+1}} \left(\frac{1}{[(t-x)^2+x_n^2]^{\frac{n-2}{2}}} \right) dt$$

$$= \frac{k(-1)^{l-\alpha_n}}{n-2} \int_{R^{n-1}} [f(t)-f(x)] \frac{\partial^{l+1}}{\partial t_1^{\alpha_1} \ldots \partial t_{n-1}^{\alpha_{n-1}} \partial x_n^{\alpha_n+1}} \left(\frac{1}{[(t-x)^2+x_n^2]^{\frac{n-2}{2}}} \right) dt.$$

We hence obtain

$$\left| \frac{\partial^l U}{\partial x_1^{\alpha_1} \ldots \partial x_n^{\alpha_n}} \right| \leqslant ckx_n^{1-l} \int_{R^{n-1}} \frac{|f(t)-f(x)|}{[(t-x)^2+x_n^2]^{\frac{n}{2}}} dt$$

$$\rho^{p(l+1)} \int_{R_\rho^{n-1}} |U^l|^p dR_\rho^{n-1} \leqslant c\rho^p \int_{R_\rho^{n-1}} \left| \int_{R^{n-1}} \frac{|f(x)-f(t)|}{[(t-x)^2+x_n^2]^{\frac{n}{2}}} dt \right|^p dx$$

$$\leqslant c \int_{R^{n-1}} \int_{R^{n-1}} \frac{|f(x+\rho t)-f(x)|^p}{[t^2+1]^{\frac{n}{2}}} dt dx \to 0 \quad (\rho \to 0).$$

Theorem 5. *Let*

$$f \in L_p(R_0^{n-1}),\ U = kx_n \int_{R^{n-1}} \frac{f}{[(t-x)^2+x_n^2]^{\frac{n}{2}}} dt.$$

Then for $p \geq 2$

$$\int_{\overline{R}_0^n} x_n^{p-1} \sum_{i=1}^n \left| \frac{\partial U}{\partial x_i} \right|^p d\overline{R}_0^n \leqslant c \| f \|^p_{L_p(R_0^{n-1})}, \tag{18}$$

where c *depends only on* p *and* n.

Proof. We shall assume that $f \geq 0$. Then $(f \not\equiv 0)$

$$U = kx_n \int_{R^{n-1}} \frac{f\,dt}{[(t-x)^2+x_n^2]^{\frac{n}{2}}} > 0.$$

It follows from (16) that

$$\left|\frac{\partial^l U}{\partial x_1^{\alpha_1}\ldots\partial x_n^{\alpha_n}}\right| \leqslant c x_n^{-l} U. \tag{19}$$

Then

$$\int_{R_\rho^n}\sum_{i=1}^n \frac{\partial U}{\partial x_i}\,\frac{\partial (U^{p-1}\cdot x_n)}{\partial x_i}\,dR_\rho^n = (p-1)\int_{\overline{R}_\rho^n}\sum_{i=1}^n\left(\frac{\partial U}{\partial x_i}\right)^2 U^{p-2}x_n d\overline{R}_\rho^n$$

$$+\int_{\overline{R}_\rho^n}\frac{\partial U}{\partial x_n}U^{p-1}d\overline{R}_\rho^n = (p-1)\int_{\overline{R}_\rho^n}\sum_{i=1}^n\left(\frac{\partial U}{\partial x_i}\right)^2 U^{p-2}x_n d\overline{R}_\rho^n - \frac{1}{p}\int_{R_\rho^{n-1}} U^p dR_\rho^{n-1}.$$

Applying Taylor's formula and taking (14) into account, we obtain

$$(p-1)\int_{\overline{R}_\rho^n}\sum_{i=1}^n\left(\frac{\partial U}{\partial x_i}\right)^2 U^{p-2}x_n dR_\rho^n - \frac{1}{p}\int_{R_\rho^{n-1}} U^p dR_\rho^{n-1}$$

$$= -\rho\int_{R_\rho^{n-1}}\frac{\partial U}{\partial x_n}U^{p-1}dR_\rho^{n-1} + \lim_{\rho_1\to\infty}\rho_1\int_{R_{\rho_1}^{n-1}}\frac{\partial U}{\partial x_n}U^{p-1}dR_{\rho_1}^{n-1}$$

$$+\lim_{\rho_1\to\infty}\lim_{N\to\infty}\sum_{i=1}^{n-1}\int_\rho^{\rho_1}\underbrace{\int_{-\infty}^{\infty}\ldots\int_{-\infty}^{\infty}}_{n-2}\frac{\partial U}{\partial x_i}(x_1,\ \ldots,\ N,\ \ldots,\ x_n)U^{p-1}x_n dx' dx_n$$

$$+\lim_{\rho_1\to\infty}\lim_{N\to\infty}\sum_{i=1}^{n-1}\int_\rho^{\rho_1}\underbrace{\int_{-\infty}^{\infty}\ldots\int_{-\infty}^{\infty}}_{n-2}\frac{\partial U}{\partial x_i}(x_1,\ \ldots,\ N,\ \ldots,\ x_n)U^{p-1}x_n dx' dx_n.$$

(Here $x' = (x_1, \cdots, x_{i-1}, x_{i+1}, \cdots, x_n)$.)

Hence, passing to the limit as $\rho \to 0$ and using estimates (17) and (19), we have

$$\int_{\overline{R}_0^n}\sum_{i=1}^n\left(\frac{\partial U}{\partial x_i}\right)^2 U^{p-2}x_n d\overline{R}_n^0 \leqslant c\lim_{\rho\to 0}\int_{R_\rho^{n-1}}|U|^p dR_\rho^{n-1} \leqslant c\int_{R_0^{n-1}}|f|^p dR_0^{n-1}.$$

$$\int_{\overline{R}_0^n}\sum_{i=1}^n\left|\frac{dU}{\partial x_i}\right|^p x_n^{p-1}d\overline{R}_0^n \leqslant c\int_{\overline{R}_0^n}\sum_{i=1}^n\left(\frac{\partial U}{\partial x_i}\right)^2 U^{p-2}x_n d\overline{R}_0^n \leqslant c\int_{R_0^{n-1}}|f|^p dR_0^{n-1}.$$

To prove the theorem in the general case, we note that

$$f = |f| - (|f| - f) = f_1 - f_2,\ f_1 \geqslant 0,\ f_2 \geqslant 0.$$

Then, setting

$$U_i = kx_n \int\limits_{R^{n-1}} \frac{f_i dt}{[(t-x)^2 + x_n^2]^{\frac{n}{2}}},\quad i = 1,\ 2,$$

we have

$$\int\limits_{\overline{R}_0^n} \sum_{i=1}^n \left|\frac{\partial U}{\partial x_i}\right|^p \cdot x_n^{p-1} d\overline{R}_0^n \leqslant c_1 \int\limits_{\overline{R}_0^n} \sum_{i=1}^n \left|\frac{\partial U_1}{\partial x_i}\right|^p x_n^{p-1} d\overline{R}_0^n$$

$$+ c_1 \int\limits_{\overline{R}_0^n} \sum_{i=1}^n \left|\frac{\partial U_2}{\partial x_i}\right|^p \cdot x_n^{p-1} d\overline{R}_0^n \leqslant c_2 \int\limits_{R_0^{n-1}} |f_1|^p dR_0^{n-1}$$

$$+ c_2 \int\limits_{R_0^{n-1}} |f_2|^p dR_0^{n-1} \leqslant c \int\limits_{R_0^{n-1}} |f|^p dR_0^{n-1}.$$

Corollary 1. *If the functions f and U satisfy the conditions of Theorem* 5, *then*

$$\int\limits_{\overline{R}_0^n} \sum_{\alpha_1+\ldots+\alpha_n=k} x_n^{kp-1} \left|\frac{\partial^k U}{\partial x_1^{\alpha_1}\ldots\partial x_n^{\alpha_n}}\right|^p d\overline{R}_0^n \leqslant c\,\|f\|^p_{L_p(R^{n-1})},$$

where c depends only on n, k and p $(p \geq 2)$.

Proof. It suffices to show that for $f \geq 0$

$$\int\limits_{\overline{R}_0^n} \sum_{\alpha_1+\ldots+\alpha_n=k} x_n^{2k-1} \left|\frac{\partial^k U}{\partial x_1^{\alpha_1}\ldots\partial x_n^{\alpha_n}}\right|^2 U^{p-2} d\overline{R}_0^n \leqslant c\,\|f\|^p_{L_p(R^{n-1})}. \tag{20}$$

The proof is by induction. Let inequality (20) hold for $l-1$; we prove it for l. By the same reasoning as above, we find that

$$\int\limits_{\overline{R}_\rho^n} \sum_{i=1}^n \frac{\partial V}{\partial x_i} \frac{\partial (VU^{p-2}x_n^{2l-1})}{\partial x_i} d\overline{R}_\rho^n = \int\limits_{\overline{R}_\rho^n} \sum_{i=1}^n \left(\frac{\partial V}{\partial x_i}\right)^2 U^{p-2} x_n^{2l-1} d\overline{R}_\rho^n$$

$$+ (p-2) \int\limits_{\overline{R}_\rho^n} \sum_{i=1}^n \frac{\partial V}{\partial x_i} V \frac{\partial u}{\partial xi} U^{p-3} x_n^{2l-1} d\overline{R}_\rho^n +$$

$$+(2l-1)\int_{\overline{R}_\rho^n}\frac{\partial V}{\partial x_n}VU^{p-2}x_n^{2(l-1)}d\overline{R}_\rho^n=\rho^{2l-1}\int_{R_\rho^{n-1}}\frac{\partial V}{\partial x_n}VU^{p-2}dR_\rho^{n-1}.$$

Hence

$$\int_{\overline{R}_\rho^n}\sum_{i=1}^n\left|\frac{\partial V}{\partial x_i}\right|^2U^{p-2}x_n^{2l-1}dR_\rho^n$$

$$\leqslant c\int_{\overline{R}_\rho^n}\sum_{i=1}^n\left|\frac{\partial V}{\partial x_i}\right||V|U^{p-2}x_n^{2(l-1)}dR_\rho^n+c\int_{R_\rho^{n-1}}U^p dR_\rho^{n-1}$$

$$\leqslant c\left[\int_{\overline{R}_\rho^n}\sum_{i=1}^n\left|\frac{\partial V}{\partial x_i}\right|^2U^{p-2}x_n^{2l-1}d\overline{R}_\rho^n\right]^{\frac{1}{2}}\left[\int_{\overline{R}_\rho^n}|V|^2U^{p-2}x_n^{2l-3}d\overline{R}_\rho^n\right]^{\frac{1}{2}}$$

$$+c\|f\|^p_{L_p(R^{n-1})}.$$

Taking into account that

$$\int_{\overline{R}_\rho^n}|V|^2U^{p-2}x_n^{2l-3}d\overline{R}_\rho^n\leqslant c\|f\|^p_{L_p(R^{n-1})},$$

we have

$$\int_{\overline{R}_0^n}\sum_{i=1}^n\left|\frac{\partial V}{\partial x_i}\right|^2U^{p-2}x_n^{2l-1}d\overline{R}_0^n$$

$$\leqslant c\left[\int_{\overline{R}_0^n}\sum_{i=1}^n\left|\frac{\partial V}{\partial x_i}\right|^2U^{p-2}x_n^{2l-1}d\overline{R}_0^n\right]^{\frac{1}{2}}\|f\|^{\frac{p}{2}}_{L_p(R^{n-1})}+c\|f\|^p_{L_p(R_0^{n-1})},$$

from which estimate (20) follows.

Corollary 2. *Let* $f\in W_p^1(R_0^{n-1})$, *where* $p\geq 2$, l *is an integer, and the harmonic function* U *satisfies the conditions of the theorem. Then*

$$\int_{\overline{R}_0^n}\sum_{\alpha_1+\ldots+\alpha_n=k}x_n^{(k-l)p-1}\left|\frac{\partial^kU}{\partial x_1^{\alpha_1}\ldots\partial x_n^{\alpha_n}}\right|^p d\overline{R}_0^n\leqslant c\|f\|^p_{W_p^l(R_0^{n-1})}, \qquad (21)$$

where the constant c does not depend on f, and $k = l + 1, \dots$.

To prove this it suffices to apply Corollary 1 to the derivatives of the function f.

Theorem 6. *Let U be a harmonic function defined on the half-space* $x_n \geq 0$, *and*

$$\int_{\overline{R}^n_0} x_n^{p-1} \sum_{i=1}^{n} \left| \frac{\partial U}{\partial x_i} \right|^p d\overline{R}^n_0 < \infty, \tag{22}$$

$$\lim_{\rho \to \infty} \int_{R^{n-1}_\rho} |U|^p dR^{n-1}_\rho < \infty. \tag{23}$$

Then if $1 < p \leq 2$, *it follows that*

$$U|_{R^{n-1}_0} = f \in L_p(R^{n-1}_0)$$

and

$$\|f\|^p_{L_p(R^{n-1}_0)} \leqslant c \int_{\overline{R}^n_0} x_n^{p-1} \sum_{i=1}^{n} \left| \frac{\partial U}{\partial x_i} \right|^p d\overline{R}^n_0, \tag{24}$$

where c does not depend on f.

Proof. It follows from (22) and (23) that for any $\rho > 0$ the harmonic function U satisfies conditions a) and b) on the half-space $x_n \geq \rho$ $(\overline{R}^n_\rho)$, and hence has the representation

$$U = k(x_n - \rho) \int_{R^{n-1}} \frac{U_\rho \, dt}{[(t-x)^2 + (x_n - \rho)^2]^{\frac{n}{2}}}, \tag{25}$$

where

$$U_\rho = U|_{R^{n-1}_\rho}.$$

We constitute the function $\phi(\rho) = |U_\rho|^{p-1} \cdot \operatorname{sign} U_\rho$. Obviously, $\phi \in L_{p'}(R^{n-1}_\rho)$, where $1/p + 1/p' = 1$ and $p' \geq 2$.

Let

$$U_1 = k(x_n - \rho) \int_{R^{n-1}} \frac{\varphi(\rho) \, dt}{[(t-x)^2 + (x_n - \rho)^2]^{\frac{n}{2}}}, \quad x_n > \rho.$$

We have

$$\int_{\overline{R}^n_\rho} \left[\sum_{i=1}^n \frac{\partial U}{\partial x_i} \cdot \frac{\partial U_1 (x_n - \rho)}{\partial x_i} + \frac{\partial U_1}{\partial x_i} \cdot \frac{\partial U(x_n - \rho)}{\partial x_i} \right] dR^n_\rho$$

$$= 2 \int_{\overline{R}^n_\rho} \sum_{i=1}^n \frac{\partial U}{\partial x_i} \cdot \frac{\partial U_1}{\partial x_i} (x_n - \rho)\, d\overline{R}^n_\rho + \int_{\overline{R}^n_\rho} \frac{\partial U_1 U}{\partial x_n} d\overline{R}^n_\rho$$

$$= \lim_{x_n \to \infty} (x_n - \rho) \int_{R^{n-1}_x} \frac{\partial U_1 U}{\partial x_n} dR^{n-1}_{x_n} - \lim_{x_n \to \rho} (x_n - \rho) \int^n_{R^{n-1}_{x_n}} \frac{\partial U U_1}{\partial x_n} dR^{n-1}_{x_n}.$$

Hence

$$\lim_{x_n \to \infty} \int_{R^{n-1}_{x_n}} U_1 U dR^{n-1}_{x_n} = \lim_{x_n \to \infty} \int_{R^{n-1}_{x_n}} U U_1 dR^{n-1}_{x_n} + 2 \int_{\overline{R}^n_\rho} \sum_{i=1}^n \frac{\partial U}{\partial x_i} \frac{\partial U_1}{\partial x_i} (x_n - \rho)\, d\overline{R}^n_\rho$$

$$+ \lim_{x_n \to \rho} (x_n - \rho) \int_{R^{n-1}_{x_n}} \frac{\partial U U_1}{\partial x_n} dR^{n-1}_{x_n} + \lim_{x_n \to \infty} (x_n - \rho) \int_{R^{n-1}_{x_n}} \frac{\partial U U_1}{\partial x_n} dR^{n-1}_{x_n}.$$

We estimate the terms on the right, applying Hölder's inequality and estimates (14), (15):

$$(x_n - \rho) \left| \int_{R^{n-1}_{x_n}} \frac{\partial U U_1}{\partial x_n} dR^{n-1}_{x_n} \right| \leqslant (x_n - \rho) \left[\left(\int_{R^{n-1}_{x_n}} \left| \frac{\partial U}{\partial x_n} \right|^p dR^{n-1}_{x_n} \right)^{\frac{1}{p}} \right.$$

$$\cdot \left(\int_{R^{n-1}_{x_n}} |U_1|^{p'} dR^{n-1}_{x_n} \right)^{\frac{1}{p'}} + \left(\int_{R^{n-1}_{x_n}} \left| \frac{\partial U_1}{\partial x_n} \right|^{p'} dR^{n-1}_{x_n} \right)^{\frac{1}{p'}}$$

$$\left. \cdot \left(\int_{R^{n-1}_{x_n}} |U|^p dR^{n-1}_{x_n} \right)^{\frac{1}{p}} \right] \leqslant (x_n - \rho) \cdot o\,[(x_n - \rho)^{-1}]$$

$$\cdot [\| U \|^{p-1}_{L_p(R^{n-1}_\rho)} + \| U_\rho \|_{L_p(R^{n-1}_\rho)}] \to 0, \quad x_n \to \infty \quad (x_n \to 0).$$

By the definition of U and U_1, as well as estimate (17), we have

$$\left| \int_{R^{n-1}_{x_n}} (U U_1 - |U_\rho|^p) dR^{n-1}_{x_n} \right| \leqslant \left| \int_{R^{n-1}_{x_n}} U [U_1 - \varphi(\rho)]\, dR^{n-1}_{x_n} + \right.$$

$$+ \int_{R^{n-1}_{x_n}} \varphi(\rho)[U-U_\rho]\, dR^{n-1}_{x_n}\Big| \leqslant \Big(\int_{R^{n-1}_{x_n}} |U_\rho|^p\, dR^{n-1}_{x_n}\Big)^{\frac{1}{p}} \cdot$$

$$\cdot \Big(\int_{R^{n-1}_{x_n}} |U_1 - \varphi(\rho)|^{p'}\, dR^{n-1}_{x_n}\Big)^{\frac{1}{p'}}$$

$$+ \Big(\int_{R^{n-1}_{x_n}} |U_\rho|^{p'}\, dR^{n-1}_{x_n}\Big)^{\frac{1}{p'}} \Big(\int_{R^{n-1}_{x_n}} |U-U_\rho|^p\, dR^{n-1}_{x_n}\Big)^{\frac{1}{p}} \to 0 \quad (x_n \to \rho).$$

Substituting these estimates in (25), we have

$$\int_{R^{n-1}_\rho} |U|^p dR^{n-1}_\rho = 2 \Big| \int_{\overline{R}^n_\rho} \sum_{i=1}^n \frac{\partial U}{\partial x_i}\frac{\partial U_1}{\partial x_i}(x_n-\rho)\, d\overline{R}^n_\rho \Big|$$

$$= 2 \Big| \int_{\overline{R}^n_\rho} \sum_{i=1}^n \frac{\partial U}{\partial x_i}\frac{\partial U_1}{\partial x_i}(x_n-\rho)^{\frac{p'-1}{p'}} (x_n-\rho)^{\frac{p-1}{p}}\, d\overline{R}^n_\rho \Big|$$

$$\leqslant c\Big(\int_{\overline{R}^n_\rho} \sum_{i=1}^n \Big|\frac{\partial U}{\partial x_i}\Big|^p (x_n-\rho)^{p-1}\, d\overline{R}^n_\rho\Big)^{\frac{1}{p}}$$

$$\cdot \Big(\int_{\overline{R}^n_\rho} \sum_{i=1}^n \Big|\frac{\partial U_1}{\partial x_i}\Big|^{p'} (x_n-\rho)^{p'-1} d\overline{R}^n_\rho\Big)^{\frac{1}{p'}}.$$

Since $p' \geq 2$, it follows by Theorem 5 that this is

$$\leqslant c_1 \Big(\int_{\overline{R}^n_\rho} \sum_{i=1}^n \Big|\frac{\partial U}{\partial x_i}\Big|^p (x_n-\rho)^{p-1} d\overline{R}^n_\rho\Big)^{\frac{1}{p}} \|(\varphi(\rho)\|_{L_{p'}(R^{n-1})}$$

$$\leqslant c_1 \Big(\int_{\overline{R}^n_\rho} \sum_{i=1}^n \Big|\frac{\partial U}{\partial x_i}\Big|^p (x_n-\rho)^{p-1} d\overline{R}^n_\rho\Big)^{\frac{1}{p}} \|U\|^{\frac{p}{p'}}_{L_p(R^{n-1}_\rho)}.$$

Hence

$$\Big(\int_{R^{n-1}_\rho} |U|^p dR^{n-1}_\rho\Big)^{\frac{1}{p}} \leqslant \Big(\int_{\overline{R}^n_\rho} \sum_{i=1}^n \Big|\frac{\partial U}{\partial x_i}\Big|^p (x_n-\rho)^{p-1} d\overline{R}^n_\rho\Big)^{\frac{1}{p}},$$

$$\lim_{\rho\to 0} \int_{R^{n-1}_\rho} |U|^p dR^{n-1}_\rho \leqslant c \int_{\overline{R}^n_0} \sum_{i=1}^n \Big|\frac{\partial U}{\partial x_i}\Big|^p x_n^{p-1} d\overline{R}^n_0.$$

We also have

$$V = kx_n \int_{R^{n-1}} \frac{U_{\rho_1} - U_{\rho_2}}{[(t-x)^2 + x_n^2]^{\frac{n}{2}}}\, dt = kx_n \int_{R^{n-1}} \frac{U_{\rho_1} dt}{[(t-x)^2 + x_n^2]^{\frac{n}{2}}}$$

$$- kx_n \int_{R^{n-1}} \frac{U_{\rho_2} dt}{[(t-x)^2 + x_n^2]^{\frac{n}{2}}} = U(x, x_n + \rho_1) - U(x, x_n + \rho_2)$$

(since U satisfies conditions a) and b) on the half-space $x_n \geq \min(\rho_1, \rho_2)$).

Then

$$\lim_{\rho \to 0} \int_{R_\rho^{n-1}} |V|^p dR_\rho^{n-1} \leqslant c \int_{\overline{R}_0^n} \sum_{i=1}^{n} \left| \frac{\partial V}{\partial x_i} \right|^p x_n^{p-1} d\overline{R}_0^n, \tag{26}$$

$$\int_{R^{n-1}} |U_{\rho_1} - U_{\rho_2}|^p dR^{n-1} \leqslant$$

$$\leqslant c \int_{\overline{R}_0^n} \sum_{i=1}^{n} \left| \frac{\partial U(x, x_n + \rho_1)}{\partial x_i} - \frac{\partial U(x, x_n + \rho_2)}{\partial x_i} \right|^p x_n^{p-1} d\overline{R}_0^n \to 0 \quad (\rho_1, \rho_2 \to 0). \tag{27}$$

From (26) and (27) we obtain the assertion of the theorem.

§ 3. Equivalence of the definitions of the classes W_p^l

We introduce a definition of classes W_p^l which is invariant relative to change of coordinates in the case of nonintegral l ($l = \overline{l} + \lambda$, $\overline{l}$ integral, $1 > \lambda > 0$).

We shall say that the function $f \in W_p^l(R^{n-1})$, if $f \in W_p^{\overline{l}}(R^{n-1})$ and for any of its $\overline{l}$th order generalized derivatives we have the relation

$$\| f^{\overline{l}} \|_{\overline{W}_p^\lambda(R^{n-1})} = \int_{R^{n-1}} \int_{R^{n-1}} \frac{|f^{\overline{l}}(x) - f^{\overline{l}}(y)|^p}{|x-y|^{n-1+p\lambda}}\, dx\, dy < \infty.$$

We set

$$\| f \|_{W_p^l(R^{n-1})} = \| f \|_{W_p^{\overline{l}}(R^{n-1})} + \sum \| f^{\overline{l}} \|_{\overline{W}_p^\lambda(R^{n-1})}. \tag{28}$$

This definition of the classes W_p^l was given in Slobodeckiĭ's articles [2,3].

We show that (28) and (3) are equivalent:

$$\int_{R^{n-1}}\int_{R^{n-1}} \frac{|f^{\bar{l}}(x)-f^{\bar{l}}(y)|^p}{|x-y|^{n-1+p\lambda}}\,dx\,dy$$

$$\leqslant c\int_{R^{n-1}}\int_{R^{n-1}} \frac{|f^{\bar{l}}(x_1,\ldots,x_{n-1})-f^{\bar{l}}(y_1,x_2,\ldots,x_{n-1})|^p}{|x-y|^{n-1+p\lambda}}\,dx\,dy+\ldots$$

$$\ldots+c\int_{R^{n-1}}\int_{R^{n-1}} \frac{|f^{\bar{l}}(y_1,\ldots,y_{n-2},x_{n-1})-f^{\bar{l}}(y)|^p}{|x-y|^{n-1+p\lambda}}\,dx\,dy$$

$$\leqslant c\sum_{j=1}^{n-1}\int_{-\infty}^{\infty}\int_{R^{n-1}} |f^{\bar{l}}(x_1,\ldots,x_{n-1})-f^{\bar{l}}(x_1,\ldots,x_{j-1},y_j,x_{j+1},\ldots,x_{n-1})|^p$$

$$\times\int_{-\infty}^{\infty}\int_{-\infty}^{\infty}\frac{dh_1\ldots dh_{n-2}\,dR^{n-1}\,dy_j}{\left[\sum_{i=1}^{n-2}h_i^2+(x_j-y_j)^2\right]^{\frac{n-1+p\lambda}{2}}}$$

$$\leqslant c\sum_{j=1}^{n-1}\int_0^{\infty}\int_{R^{n-1}}\frac{|f^{\bar{l}}(x_1,\ldots,x_j+h,\ldots,x_{n-1})-f^{\bar{l}}(x)|^p}{h^{1+p\lambda}}\,dR^{n-1}\,dh.$$

To prove the assertion it suffices to show that the norm (3) is estimated in terms of the norm (28).

We consider the function

$$U=kx_n\int_{R^{n-1}}\frac{f(x)}{[(t-x)^2+x_n^2]^{\frac{n}{2}}}\,dt.$$

We snow that

$$\|U\|_{L^{\bar{l}+1}_{p,\,p(1-\lambda)-1}(\bar{R}^n_0)}\leqslant c\,\|f\|_{W^l_p(R^{n-1})}, \tag{29}$$

where the norm on the right is taken in the sense of (28).

Without loss of generality, we need only estimate the derivatives with respect to $x_1, \dots, x_{n-1}$.

For $\nu \geq \bar{l} + 2$ $(\nu - \bar{l} = \beta_1 + \dots + \beta_{n-1})$ we have

$$\frac{\partial^\alpha U}{\partial x_1^{\alpha_1} \dots \partial x_{n-1}^{\alpha_{n-1}}} = (-1)^{\nu-\bar{l}} k x_n \int_{R^{n-1}} f^{\bar{l}} \frac{\partial^{\nu-\bar{l}}}{\partial t_1^{\beta_1} \dots \partial t_{n-1}^{\beta_{n-1}}} \left(\frac{1}{[(t-x)^2 + x_n^2]^{\frac{n}{2}}} \right) dt$$

$$= (-1)^{\nu-\bar{l}} k x_n \int_{R^{n-1}} [f^{\bar{l}}(x+t) - f^{\bar{l}}(x)] \frac{\partial^{\nu-\bar{l}}}{\partial t_1^{\beta_1} \dots \partial t_{n-1}^{\beta_{n-1}}} \left(\frac{1}{[(t-x)^2 + x_n^2]^{\frac{n}{2}}} \right) dt,$$

$$\left| \frac{\partial^\nu U}{\partial x_1^{\alpha_1} \dots \partial x_{n-1}^{\alpha_{n-1}}} \right| \leqslant c \int_{R^{n-1}} \frac{|\Delta f^{\bar{l}}| \, dt}{[(t-x)^2 + x_n^2]^{\frac{n+\nu-\bar{l}-1}{2}}},$$

$$\int_{\bar{R}_0^n} x_n^{(\nu-l)p-1} |U^\nu|^p d\bar{R}_0^n \leqslant c \int_{\bar{R}_0^n} x_n^{(\nu-l)p-1} \left| \int_{R^{n-1}} \frac{|\Delta f^{\bar{l}}| \, dt}{[(t-x)^2 + x_n^2]^{\frac{n+\nu-\bar{l}-1}{2}}} \right|^p d\bar{R}_0^n$$

$$\leqslant c \int_{\bar{R}_0^n} x_n^{(\nu-l)p-1} \int_{R^{n-1}} \frac{|\Delta f^{\bar{l}}|^p dt}{[(t-x)^2 + x_n^2]^{\frac{\varepsilon p}{2}}} \left[\int_{R^{n-1}} \frac{dt}{[(t-x)^2 + x_n^2]^{\frac{(n+\nu-\bar{l}-1-\varepsilon)q}{2}}} \right]^{\frac{p}{q}} d\bar{R}_0^n \quad ^{1)}$$

$$\leqslant c \int_{R^{n-1}} \int_{R^{n-1}} |\Delta f^{\bar{l}}|^p \int_0^\infty \frac{x_n^{-(n+p\lambda)+\varepsilon p}}{[(t-x)^2 + x_n^2]^{\frac{\varepsilon p}{2}}} dx_n dt \, dx$$

$$\leqslant c \int_{R^{n-1}} \int_{R^{n-1}} \frac{|\Delta f^l|^p dt \, dx}{(t-x)^{n-1+p\lambda}} \int_0^\infty \frac{u^{-(n+p\lambda)+\varepsilon p}}{[1+u^2]^{\frac{\varepsilon p}{2}}} du \leqslant c \|f\|_{W_p^l(R^{n-1})}.$$

This is arrived at by substituting $x_n/|t-x| = u$ $(-n - p\lambda + \epsilon p > -1)$.

---

1) Here

$$\frac{n-1}{p} + \lambda < \varepsilon < \frac{n}{p} + \lambda, \quad \frac{1}{p} + \frac{1}{q} = 1,$$

$$(n + \nu - \bar{l} - 1 - \varepsilon) q > (n - \varepsilon + \lambda) q > n \left(1 - \frac{1}{p}\right) q = n$$

By (1.4) and the Corollary to Theorem 1, this implies (29).

We introduce the function F:

$$F\equiv U,\quad 0\leqslant x_n\leqslant 1;$$

$$F\equiv 0,\qquad x_n\geqslant 2;$$

$$F=Ue^{-\left[\frac{x_n-1}{x_n-2}\right]^{2l}},\quad 1\leqslant x_n\leqslant 2. \tag{30}$$

Obviously $F\in W^{\bar{l}+1}_{p,p(1-\lambda)-1}(\bar{R}^n_0)$ and

$$\|F\|_{W^{\bar{l}+1}_{p},\,p(1-\lambda)-1}(\bar{R}^n_0)\leqslant c\|f\|_{W^l_p(R^{n-1})},$$

where the norm on the right is taken in the sense of (28).

To complete the proof, it suffices to apply Theorem 2 to the function f.

The following lemma generalizes results in [15,16,6].

Lemma 3. *Let f be defined on $\bar{R}^n_0$ and belong to the class $W^l_p(\bar{R}^n_0)$ (l is any positive number). Then there exists a function $F\in W^l_p(R^n)$, such that*

$$F\equiv f\ \text{on}\ \bar{R}^n_0$$

and

$$\|F\|_{W^l_p(R^n)}\leqslant c\|f\|_{W^l_p(\bar{R}^n_0)}, \tag{31}$$

where the constant c does not depend on f.

Proof. We construct the function F by the method of Whitney and Hestenes (see, for example, [17]).

Let $l=\bar{l}+\lambda$; we define constants β_m by the system of equations

$$\sum_{m=0}^{\bar{l}}\left(-\frac{1}{m+1}\right)^k\cdot\beta_m=1,\quad k=0,1,\ldots,\bar{l}.$$

We set

$$F=\begin{cases} f & \text{on } \bar{R}^n_0,\\ \bar{f}=\sum\limits_{m=0}^{\bar{l}}\beta_m f\left(x,-\dfrac{x_n}{m+1}\right) & \text{on } R^n-\bar{R}^n_0.\end{cases}$$

It is easily seen that the function F has all generalized derivatives through order $\bar{l}$, inclusive, and

$$\|F\|_{W_p^{\bar{l}}(R^n)} \leqslant c\|f\|_{W_p^{\bar{l}}(\bar{R}_0^n)}$$

(see, for example, [6], Russian p. 69).

To complete the proof of the lemma, it suffices to show the inequality

$$\|F^{\bar{l}}\|_{\bar{W}_p^\lambda(R^n)} \leqslant c\|f^{\bar{l}}\|_{\bar{W}_p^\lambda(\bar{R}_0^n)},$$

where

$$F^{\bar{l}} = \frac{\partial^{\bar{l}} F}{\partial x_1^{\alpha_1} \ldots \partial x_n^{\alpha_n}}, \quad f^{\bar{l}} = \frac{\partial^{\bar{l}} f}{\partial x^{\alpha_1} \ldots \partial x_n^{\alpha_n}}, \qquad \alpha_1 + \ldots + \alpha_n = \bar{l}.$$

We have $(l_1 \leq \bar{l})$

$$\int_0^\infty \int_{R^n} \frac{|F^{\bar{l}}(x_1, \ldots, x_n + h) - F^{\bar{l}}(x_1, \ldots, x_n)|^p}{h^{1+p\lambda}} dR^n dh$$

$$\leqslant \int_0^\infty \int_{-h}^0 \int_{R^{n-1}} \frac{|F^{\bar{l}}(x_1, \ldots, x_n + h) - F^{\bar{l}}(x_1, \ldots, x_n)|^p}{h^{1+p\lambda}} dR^{n-1} dx_n dh$$

$$+ \sum_{m=0}^{\bar{l}} \beta_m \left(-\frac{1}{m+1}\right)^{l_1}$$

$$\cdot \int_0^\infty \int_{-\infty}^h \int_{R^{n-1}} \frac{\left|f^{\bar{l}}\left(x_1, \ldots, -\frac{x_n+h}{m+1}\right) - f^{\bar{l}}\left(x_1, \ldots, -\frac{x_n}{m+1}\right)\right|^p}{h^{1+p\lambda}} dR^{n-1} dh\, dx_n$$

$$+ \|f^{\bar{l}}\|^p_{\bar{W}_p^\lambda(\bar{R}_0^n)} \leqslant \int_{-\infty}^0 \int_{-x_n}^\infty \int_{R^{n-1}} \frac{|F^{l}(x_1, \ldots, x_n + h) - F^{\bar{l}}(x_1, \ldots, x_n)|^p}{h^{1+p\lambda}} dR^{n-1} dh\, dx_n$$

$$+ c\|f^{\bar{l}}\|^p_{\bar{W}_p^\lambda} \leqslant \int_{-\infty}^0 \int_0^\infty \int_{R^{n-1}} \frac{|F^{\bar{l}}(x_1, \ldots, x_{n-1}, h) - F^{\bar{l}}(x_1, \ldots x_n)|^p}{|h - x_n|^{1+p\lambda}} dR^{n-1} dh\, dx_n$$

$$+ c\|f^{\bar{l}}\|^p_{\bar{W}_p^\lambda} \leqslant c\|f^{\bar{l}}\|^p_{\bar{W}_p^\lambda} +$$

$$+\int_{-\infty}^{0}\int_{0}^{\infty}\int_{R^{n-1}} \frac{\left|f^{\bar{l}}(x_1,\ldots,x_{n-1}h)-\sum_{m=0}^{\bar{l}}\beta_m\left(-\frac{1}{m+1}\right)^{l_1} f^{\bar{l}}\left(x_1,\ldots,-\frac{x_n}{m+1}\right)\right|^p}{|h-x_n|^{1+p\lambda}}\, dR^{n-1}dhdx_n$$

$$\leqslant c\|f^{\bar{l}}\|^p_{\overline{W}^{\lambda}_p}+c\int_0^{\infty}\int_{\bar{R}^n_0}\frac{|f^{\bar{l}}(x_1,\ldots,x_{n-1},h)-f^{\bar{l}}(x_1,\ldots,x_n)|^p}{|h-x_n|^{1+p\lambda}}\,d\bar{R}^n_0dh$$

$$\leqslant c\|f^{\bar{l}}\|^p_{\overline{W}^{\lambda}_p}+c\int_0^{\infty}\int_{-x_n}^{0}\int_{R^{n-1}}\frac{|f^{\bar{l}}(x_1,\ldots,x_n+h)-f^{\bar{l}}(x_1,\ldots,x_n)|^p}{|h|^{1+p\lambda}}\,dR^{n-1}dh\,dx_n$$

$$\leqslant c\|f^{\bar{l}}\|^p_{\overline{W}^{\lambda}_p}+c\int_0^{\infty}\int_h^{\infty}\int_{R^{n-1}}\frac{|f^{\bar{l}}(x_1,\ldots,x_n-h)-f^{\bar{l}}(x_1,\ldots,x_n)|^p}{h^{1+p\lambda}}\,dR^{n-1}dx_n\,dh$$

$$\leqslant c\|f^{\bar{l}}\|^p_{\overline{W}^{\lambda}_p(\bar{R}^n_0)}.$$

Lemma 4. *If l is an integer, then*

$$W^l_p(R^{n-1})\subset B^l_p(R^{n-1}) \quad \textit{for} \quad p\geqslant 2,$$
$$B^l_p(R^{n-1})\subset W^l_p(R^{n-1}).$$

For $n=1$ this is a result of Marcinkiewicz [18]. As Besov showed, this result can be generalized to $n\geq 1$. Below we give a proof based on other methods.

Proof. Let $f\in W^r_p(R^{n-1})$; then, by Corollary 2 of Theorem 5, the harmonic function U satisfying conditions a) and b) also satisfies the inequality

$$\|U\|_{L^l_{p,p(l-r)-1}(\bar{R}^n_0)}\leqslant c\|f\|_{W^r_p(R^{n-1})},$$

where $l\geq r+1$ and c does not depend on f.

As above, we can construct a function such that $F\in W^l_{p,p(l-r)-1}$,

$$F|_{R^{n-1}}=f,$$

$$\|F\|_{W^l_{p,p(l-r)-1}(\bar{R}^n_0)}\leqslant c\|f\|_{W^r_p(R^{n-1})},$$

where c does not depend on f.

By Theorem 2, this means that for $p \geq 2$

$$\|f\|_{B_p^r} \leqslant c\|F\|_{W^l_{p,p(l-r)-1}(\overline{R}_0^n)} \leqslant c_1\|f\|_{W_p^r}.$$

If $f \in B_p^r(R^{n-1})$, $1 < p \leq 2$, then for the harmonic function U which satisfies conditions a) and b) we have, by Theorem 3 [12],

$$\|U\|_{L^{r+1}_{p,p-1}(\overline{R}_0^n)} \leqslant c\|f\|_{B_p^r(R^{n-1})}.$$

Let

$$U^r = \frac{\partial^r U}{\partial x_1^{\alpha_1} \ldots \partial x_{n-1}^{\alpha_{n-1}}}.$$

Then, by Theorem 6, for $1 < p \leq 2$ we have

$$U^r\big|_{R^{n-1}} = f^r \in L_p(R^{n-1})$$

and

$$\|f^r\|_{L_p(R^{n-1})} \leqslant c\|U^r\|_{L^1_{p,p-1}(\overline{R}_0^n)} \leqslant c_1\|f\|_{B_p^r(R^{n-1})},$$

or

$$\|f\|_{W_p^r(R^{n-1})} \leqslant c\|f\|_{B_p^r(R^{n-1})}.$$

§ 4. Boundary values of the classes $W_p^l(R^{n-1})$

For $p = 2$ the following theorem is due to Slobodeckiĭ [6], and for $p \neq 2$ and l or r not an integer it is due to Besov [9].

Theorem 7. *Let* $1 \leq m < n-1$, $p > 1$ *and* $r = l - (n-m-1)/p > 0$. *Then*

$$W_p^l(R^{n-1}) \subset B_p^r(R^m).$$

Proof. Let $f \in W_p^l(R^{n-1})$; taking Theorem 2 into account, we may assume that l is not an integer. By Theorem 3, we may then construct a function of the form (30) such that

$$V|_{R^{n-1}} = f,$$

$$V \in W^{\nu}_{p,p(\nu-l)-1}(\bar{R}^n_0), \quad \nu = \bar{l}+1, \ldots,$$

where

$$\|V\|_{W^{\nu}_{p,p(\nu-l)-1}(\bar{R}^n_0)} \leqslant c\|f\|_{W^l_p(R^{n-1})}, \tag{32}$$

and c does not depend on f or V.

We compose the function

$$F = kx_n \int\limits_{R^{n-1}} \frac{V(t, x_n)}{[(t-x)^2 + x_n^2]^{\frac{n}{2}}} dt. \tag{33}$$

We show that $F \in W^{\nu}_{p,p(\nu-r)-1}(\bar{R}^{m+1}_0)$, and

$$\|F\|_{W^{\nu}_{p,p(\nu-r)-1}(\bar{R}^{m+1}_0)} \leqslant c\|f\|_{W^l_p(R^{n+1})},$$

where c does not depend on F or f and

$$\bar{R}^{m+1}_0 = [(x_1, \ldots, x_m) \in R^m, \quad x_n \geqslant 0].$$

We have

$$F^{\nu}_{\alpha_1\ldots\alpha_n} = k \int\limits_{R^{n-1}} \sum_{i=1}^{\alpha_n} C_i^{\alpha_n} \frac{\partial^i V^{\nu-\alpha_n}_{\alpha_1\ldots\alpha_{n-1}}}{\partial x_n^i} \frac{\partial^{\alpha_n - i}}{\partial x_n^{\alpha_n - i}} \left(\frac{x_n}{[(t-x)^2 + x_n]^{\frac{n}{2}}} \right) dt$$

$$\left(f^{\nu}_{\alpha_1\ldots\alpha_n} = \frac{\partial^{\nu} f}{\partial x_1^{\alpha_1} \ldots \partial x_n^{\alpha_n}} \right).$$

Hence, since the function $x_n/[(t-x)^2 + x_n^2]^{n/2}$ $(x_n > 0)$ is harmonic, we obtain

$$|F^{\nu}_{\alpha_1\ldots\alpha_n}| \leqslant \sum_{\alpha_1+\ldots+\alpha_n=\nu} c \int\limits_{R^{n-1}} |V^{\nu}_{\alpha_1\ldots\alpha_n}| \frac{x_n}{[(t-x)^2 + x_n^2]^{\frac{n}{2}}} dt +$$

$$+\sum_{\alpha_1+\ldots+\alpha_n=\nu-1} c\int_{R^{n-1}} |V^{\nu-1}_{\alpha_1\ldots\alpha_n}| \frac{1}{[(t-x)^2+x_n^2]^{\frac{n}{2}}}\,dt.$$

Since the estimates are analogous, it is sufficient to examine one term of sum (34).

Let $x^0_{m+1}, \ldots, x^0_{n-1}$ be arbitrary, but fixed. We have

$$\int_0^\infty \int_{R^m} x_n^{p(\nu-r)-1} |F^\nu_{\alpha_1\ldots\alpha_n}(x_1, \ldots, x_m, x^0_{m+1}, \ldots, x^0_{n-1}, x_n)|^p\, dR^m dx_n$$

$$\leqslant c\int_0^\infty \int_{R^m} x_n^{p(\nu-r)} \int_{R^{n-1}} \frac{|V^\nu_{\alpha_1\ldots\alpha_n}(t, x_n)|^p\, dt\, dR^m dx_n}{\left[\sum_{i=1}^m (t_i-x_i)^2+\sum_{i=m+1}^{n-1}(t_i-x_i^0)^2+x_n^2\right]^{\frac{n}{2}}}$$

$$\leqslant c\int_0^\infty x_n^{p(\nu-r)} \int_{R^{n-1}} \frac{|V^\nu_{\alpha_1\ldots\alpha_n}(t, x_n)|^p\, dt\, dx_n}{\left[\sum_{i=m+1}^{n-1}(t_i-x_i^0)^2+x_n^2\right]^{\frac{n-m}{2}}}$$

$$\leqslant c\int_0^\infty x_n^{p(\nu-r)+m-n} \int_{R^{n-1}} |V^\nu_{\alpha_1\ldots\alpha_n}(t, x_n)|^p\, dt\, dx_n \text{ [1]}$$

$$= c\int_{\overline{R}^n_0} x_n^{p(\nu-l)-1} |V^\nu_{\alpha_1\ldots\alpha_n}|\, d\overline{R}^n_0 \leqslant c\,\|f\|^p_{W^l_p(R^{n-1})}.$$

Taking into account that V has the form (30), we find from the Corollary to Theorem 1 that

$$\|F\|_{L_p\left(\overline{R}^{m+1}_0\right)} \leqslant c\,\|F\|_{W^\nu_{p,p(\nu-r)-1}\left(\overline{R}^{m+1}_0\right)} \leqslant c_1\,\|f\|_{W^l_p(R^{n-1})}.$$

If follows from Theorem 2 that

$$F|_{R^m} = \varphi \in B^r_p(R^m)$$

and

$$\|\varphi\|_{B^r_p(R^m)} \leqslant c\,\|f\|_{W^l_p(R^{n-1})}, \tag{34}$$

---

1) Since $r = l-(n-1-m)/p$, we have $p(\nu-r)+m-n = p\nu - pl + n - 1 - m + m - n = p(\nu-l)-1$.

where c does not depend on ϕ, f or $x^0_{m+1}, \cdots, x^0_{n-1}$.

To complete the proof, it suffices to show that

$$f\,|_{R^m} = \varphi. \tag{35}$$

It follows from the proof of Lemma 3.13 of article [21] and a theorem of Slobodeckiĭ [19] that there exists a sequence of continuous functions (f^k) such that, uniformly in $x^0_{m+1}, \cdots, x^0_{n-1}$,

$$\lim_{k\to\infty} \int_{R^m} |f - f^k|^p \, dR^m \quad 0. \tag{36}$$

For sufficiently small $x_n > 0$ we have

$$\int_{R^m} |\varphi - f|^p \, dR^m \leqslant c \left[\int_{R^m} |\varphi - F|^p \, dR^m + \int_{R^m} |f - F|^p \, dR^m \right] \tag{37}$$

$$\int_{R^m} |f - F|^p \, dR^m \leqslant \int_{R^m} \left[k x_n \int_{R^{n-1}} \frac{|v(t, x_n) - f(x)|}{[(t-x)^2 + x_n^2]^{\frac{n}{2}}} \, dt \right]^p dR^m$$

$$\leqslant c x_n \left[\int_{R^m} \int_{R^{n-1}} \frac{|v(t, x_n) - f^k(t)|^p}{[(t-x)^2 + x_n^2]^{\frac{n}{2}}} \, dt \, dR^m \right.$$

$$\left. + \int_{R^m} \int_{R^{n-1}} \frac{|f^k(t) - f(x)|^p + |f(x) - f^k(x)|^p}{[(t-x)^2 + x_n^2]^{\frac{n}{2}}} \, dt \, dR^m \right]$$

$$\leqslant c \int_{R^{n-1}} \int_{R^{n-1}} \int_{R^m} \left(\frac{|f[x + x_n(u+t)] - f^k[x + x_n(u+t)]|^p}{[t^2+1]^{\frac{n}{2}} [u^2+1]^{\frac{n}{2}}} \right.$$

$$\left. + \frac{|f^k[x + x_n(u+t)] - f^k(x + x_n t)|^p}{[t^2+1]^{\frac{n}{2}} [u^2+1]^{\frac{n}{2}}} \right) dR^m \, dt \, du$$

$$+ \int_{R^{n-1}} \int_{R^m} \frac{|f^k(x + x_n t) - f^k(x)|^p}{[t^2+1]^{\frac{n}{2}}} \, dR^m \, dt + \int_{R^m} |f(x) - f^k(x)|^p \, dR^m \to 0$$

$$(x_n \to 0, \ k \to \infty),$$

taking (36) into consideration.

Hence, by (37) and (35), we obtain

$$f|_{R^m} = \varphi.$$

Corollary 1. *Suppose that the conditions of Theorem 7 are fulfilled, and, in addition,* $1 < p \leq 2$. *Then*

$$W_p^l(R^{n-1}) \subset W_p^r(R^m).$$

This follows from Theorem 7 and Lemma 3.

For integral l and r this is a refinement of results of Sobolev and Kondrašov [1].

Corollary 2. *Let* $1 \leq m - n$, $p > 1$ *and* $r = l - \alpha/p - (n - m)/p > 0$. *Then*

$$W_{p,\alpha}^l(\bar{R}_0^n) \subset B_p^r(R^m).$$

Corollary 2 strengthens a result of Kudrjavcev ([22], p. 100).

Proof. If $m = n - 1$, this is Theorem 2.

Let $f \in W_{p,\alpha}^l$ and $f|_{R^{n-1}} = \phi$. It then follows from Theorems 2 and 3 that there exists a function v of the form (30), which is defined on $\bar{R}_0^n$, such that

$$v|_{R^{n-1}} = \varphi, \ v \in W_{p,\alpha}^l(\bar{R}_0^n)$$

and

$$\|v\|_{W_{p,\alpha}^l(\bar{R}_0^n)} \leqslant c\|f\|_{W_{p,\alpha}^l(\bar{R}_0^n)},$$

where c does not depend on the functions f and v.

It now suffices to apply the method of proof of Theorem 7.

For $p = 2$ the following theorem is due to Slobodeckiĭ [6], and for $p \neq 2$ and l or r integral it is due to Besov [9].

Theorem 8 (converse). *Let*

$$f \in B_p^r(R^m), \quad 1 \leqslant m < n, \ p > 1, \quad l = r + \frac{n-m}{p}.$$

Then there exists a function $v \in W_p^l(R^n)$ *such that*

$$v|_{R^m} = f,$$

$$\|v\|_{W_p^l(R^n)} \leqslant c\|f\|_{B_p^r(R^m)}, ^{1)}$$

where c does not depend on v or f.

The assertion of the theorem follows from Theorems 3, 4 and 7.

Corollary. *Let* $f \in W_p^r(R^m)$, $r \geq 0$, $1 \leq m < n$, $p \geq 2$ *and* $l = r + (n-m)/p$. *Then there exists a function* $v \in W_p^l(R^n)$ *such that*

$$v|_{R^m} = f,$$

$$\|v\|_{W_p^l(R^n)} \leqslant c\|f\|_{W_p^r(R^m)}.$$

This follows from Theorems 7 and 8 and Lemma 3.

§5. Fundamental theorem on imbedding the classes W_p^l

Theorem 9. *Let* $1 \leq m \leq n-1$, $1 < p < p' < \infty$,

$$r = l - \frac{(n-m-1)}{p} - \left(\frac{1}{p} - \frac{1}{p'}\right)m. \tag{38}$$

Then

$$W_p^l(R^{n-1}) \subset W_p^r(R^m).$$

For r and l integral, these are results of Sobolev [25] and Il'in [24].

Proof. Let $f \in W_p^l(R^{n-1})$. We consider three cases.

a) l and r are nonintegral.

We construct the function

$$F = kx_n \int\limits_{R^{n-1}} \frac{v(t, x_n)}{[(t-x)^2 + x_n^2]^{\frac{n}{2}}} dt,$$

where $v(t, x_n)$ satisfies condition (32).

To prove the theorem in our case it is then sufficient to show that

$$\|F\|_{W^{\nu}_{p', p'(\nu - r) - 1}(\overline{R}_0^{m+1})} \leqslant c\|f\|_{W_p^l(R^{n-1})}, \tag{39}$$

where c does not depend on F or f.

It follows from Theorem 1 (see [19]) for $\nu > \bar{l} + 1$ that

1) We note that for nonintegral r Theorem 8 was obtained independently and simultaneously by P. I. Lizorkin.

$$\rho^{p(\nu-l)} \int\limits_{R_\rho^{n-1}} \left| \frac{\partial^\nu v(t, x_n)}{\partial x_n^{\alpha_n} \partial t_1^{\alpha_1} \ldots \partial t_{n-1}^{\alpha_{n-1}}} \right|^p dR_\rho^{n-1} \leqslant c \| f \|_{W_p^l}^p . \tag{40}$$

For brevity, let

$$f^\nu(x, x_n) = \frac{\partial^\nu f(x, x_n)}{\partial x_1^{\alpha_1} \ldots \partial x_n^{\alpha_n}}.$$

Taking (34) into account and limiting ourselves to estimating only one term of the sum on the right in (34), we have

$$\int\limits_{\overline{R}_0^{m+1}} x_n^{p'(\nu-r)-1} | F^\nu(x, x_n) |^{p'} d\overline{R}_0^{m+1}$$

$$\leqslant c \int\limits_{\overline{R}_0^{m+1}} x_n^{p'(\nu-r)+\frac{p'}{p}-1} \left[\int\limits_{R^{n-1}} \frac{| v^\nu(t, x_n) |^p dt}{\left[\sum\limits_{i=1}^m (t_i - x_i)^2 + \sum\limits_{i=m+1}^{n-1} (t_i - x_i^0)^2 + x_n^2 \right]^{\frac{n}{2}}} \right]^{\frac{p'}{p}} d\overline{R}_0^m$$

$$\leqslant c \int\limits_{\overline{R}_0^{m+1}} x_n^{p'(\nu-r)+\frac{p'}{p}-1} \int\limits_{R^{n-1}} \frac{| v^\nu(t, x_n) |^p x_n^{-\frac{n(p'-p)}{p}} dt}{\left[\sum\limits_{i=1}^m (t_i - x_i)^2 + x_n^2 \right]^{\frac{n}{2}}}$$

$$\times \left[\int\limits_{R^{n-1}} \frac{| v^\nu(t, x_n) |^p}{} dt \right]^{\frac{p'-p}{p}} d\overline{R}_0^{m+1}.$$

Taking (40) into account and setting $\beta = p'(l-r) + p(\nu - l) - (n-1)(1 - p'/p)$, we find that

$$\int\limits_{\overline{R}_0^{m+1}} x_n^{p'(\nu-r)-1} | F^\nu(x, x_n) |^{p'} d\overline{R}_0^{m+1}$$

$$\leqslant c \| f \|_{W_p^l(R^{n-1})}^{p'-p} \int\limits_0^\infty x_n^{\beta+m-n} \int\limits_{R^{n-1}} | v^\nu(t, x_n) |^p dR^{n-1} dx_n.$$

Using (38) and the fact that $p - n + m = p(\nu - l) - 1$, we have

$$\int\limits_{\overline{R}_0^{m+1}} x_n^{p'(\nu-r)-1} |F^\nu(x, x_n)|^{p'} d\overline{R}_0^{m+1}$$

$$\leqslant c\|f\|_{W_p^l(R^{n-1})}^{p'-p} \int\limits_{\overline{R}_0^n} x_n^{p(\nu-l)-1} |v^\nu(x, x_n)|^p d\overline{R}_0^n \leqslant c\|f\|_{W_p^l(R^{n-1})}^{p'}.$$

Since by construction $F \equiv 0$ in the region $x_n \geq 2$, it follows from the Corollary to Theorem 1 that

$$\|F\|_{L_{p'}(\overline{R}_0^{m+1})} \leqslant c\|F\|_{W^\nu_{p', p'(\nu-r)-1}(\overline{R}_0^{m+1})} \leqslant c_1 \|f\|_{W_p^l}.$$

(39) is thus completely proved.

b) Let l be an integer and r not an integer.

Taking a) into account, without loss of generality we may assume that $m = n - 1$.

Let $\phi \in L_p(R^{n-1})$, $q > p$ and

$$v = \int\limits_{R^{n-1}} \frac{\varphi(t)}{[(t-x)^2 + x_n^2]^{\left(\frac{n-1}{p'} + \frac{n}{q}\right)\frac{1}{2}}} dt,$$

where

$$\frac{1}{p} + \frac{1}{p'} = 1.$$

We show that $v \in L_q(R^n)$ and

$$\|v\|_{L_q(R^n)} \leqslant c\|\varphi\|_{L_p(R^{n-1})}, \tag{41}$$

where c does not depend on v or ϕ.

It obviously suffices to show that the functional

$$I[f] = \int\limits_{R^n} vf\, dR^n$$

is defined on the entire space $L_{q'}(R^n)$ $(1/q + 1/q' = 1)$ and

$$|I[f]| \leqslant c\|v\|_{L_p(R^{n-1})} \|f\|_{L_{q'}(R^n)}. \tag{42}$$

We state a lemma that was proved by Il′ in [24].

Lemma. *Let* $\phi \in L_p(R^m)$ *and* $G \in L_{q'}(R^m)$, *where* $p > 1$, $q' > 1$, $1/p + 1/q' > 1$, $m \leq n$. *Then*

$$\int\limits_{R^n}\int\limits_{R^m} \frac{G(x_1, \ldots, x_n)\,\varphi(t_1, \ldots, t_m)}{\left[\sum\limits_{i=1}^{m}(t_i - x_i)^2 + \sum\limits_{i=m+1}^{n} x_i^2\right]^{\lambda}}\,dt\,dx \leqslant c\,\|\varphi\|_{L_p}\|G\|_{L_{q'}}. \tag{43}$$

Here

$$\lambda = \frac{m}{p'} + \frac{n}{q}, \quad \frac{1}{p} + \frac{1}{p'} = 1, \quad \frac{1}{q} + \frac{1}{q'} = 1,$$

and the constant c does not depend on ϕ or G.

Since

$$I[G] = \int\limits_{R^n}\int\limits_{R^{n-1}} \frac{\varphi(t_1, \ldots, t_{n-1})\cdot G(x_1, \ldots, x_n)}{\left[\sum\limits_{i=1}^{n-1}(t_i - x_i)^2 + x_n^2\right]^{\left(\frac{n-1}{p'} + \frac{n}{q}\right)\cdot\frac{1}{2}}}\,dt\,dx$$

and $1/p + 1/q' > 1$, it follows by Il'in's lemma that

$$|I[G]| \leqslant c\,\|G\|_{L_{q'}(R^n)}\|\varphi\|_{L_p(R^{n-1})}.$$

From this we obtain (41).

We consider the harmonic function

$$U = kx_n \int\limits_{R^{n-1}} \frac{f(t)}{[(t-x)^2 + x_n^2]^{\frac{n}{2}}}\,dt, \quad f \in W_p^l(R^{n-1}).$$

For $\nu \geq l + 2$ we have

$$\left|\frac{\partial^\nu U}{\partial x_1^{\alpha_1}\ldots\partial x_n^{\alpha_n}}\right| \leqslant c \sum_{\alpha_1+\ldots+\alpha_{n-1}=l}\ \int\limits_{R^{n-1}} \left|\frac{\partial^l f}{\partial t_1^{\alpha_1}\ldots\partial t_{n-1}^{\alpha_{n-1}}}\right| \frac{dt}{[(t-x)^2 + x_n^2]^{\frac{n+\nu-l-1}{2}}}. \tag{44}$$

Estimating only one term of the sum (44), we have

$$\int\limits_{\overline{R}_0^n} x_n^{q(\nu-r)-1}\left|\frac{\partial^\nu U}{\partial x_1^{\alpha_1}\ldots\partial x_n^{\alpha_n}}\right|^q d\overline{R}_0^n$$

$$\leqslant c\int\limits_{\overline{R}_0^n}\left[\int\limits_{R^{n-1}}\left|\frac{\partial^l f}{\partial t_1^{\alpha_1}\ldots\partial t_{n-1}^{\alpha_{n-1}}}\right|\frac{dt\cdot x_n^{\nu-r-\frac{1}{q}}}{[(t-x)^2+x_n^2]^{\frac{n+\nu-l-1}{2}}}\right]^q d\overline{R}_0^n \leqslant$$

$$\leqslant c \int_{\overline{R}^n} \left[\int_{R^{n-1}} \left| \frac{\partial^l f}{\partial t_1^{\alpha_1} \ldots \partial t_{n-1}^{\alpha_{n-1}}} \right| \frac{dt}{[(t-x)^2 + x_n^2]^{\left(\frac{n-1}{p'} + \frac{n}{q}\right)\frac{1}{2}}} \right]^q d\overline{R}_0^n$$

$$\leqslant c \sum \left\| \frac{\partial^l f}{\partial t_1^{\alpha_1} \ldots \partial t_{n-1}^{\alpha_{n-1}}} \right\|_{L_p(R^{n-1})}^q ,$$

by (41).

Hence, by the Corollary to Theorem 1 and by Lemma 2,

$$\int_{\overline{R}_0^n} x_n^{q(\overline{r}-r+1)-1} \left| \frac{\partial^{\overline{r}+1} U}{\partial x_1^{\alpha_1} \ldots \partial x_n^{\alpha_n}} \right|^q d\overline{R}_0^n \leqslant c \, \| f \|_{W_p^l} .$$

We further have for $x_n = 2$, $k = 0, 1, \cdots, \overline{r}$

$$\int_{R_2^n} \left| \frac{\partial^k U}{\partial x_1^{\alpha_1} \ldots \partial x_n^{\alpha_n}} \right|^q dR^{n-1} \leqslant c \int_{R^{n-1}} \left[\int_{R^{n-1}} \frac{|f| \, dt}{[(t-x)^2 + 2]^{\frac{n}{2}}} \right]^q dR^{n-1}$$

$$\leqslant c \int_{R^{n-1}} \int_{R^{n-1}} \frac{|f|^p \, dt \, dx}{[(t-x)^2 + 2]^{\frac{n}{2}}} \left[\int_{R^{n-1}} |f|^p dR^{n-1} \right]^{\frac{q}{p}-1} \leqslant c \, \| f \|_{L_p(R^{n-1})}^{\frac{q}{p}} . \tag{45}$$

Then, taking (45) into account, we obtain that

$$\int_0^2 \int_{R^{n-1}} \left| \frac{\partial^k U}{\partial x_1^{\alpha_1} \ldots \partial x_n^{\alpha_n}} \right|^q dR^{n-1} dx_n \leqslant c \, \| f \|^q W_p^l (R^{n-1}), \qquad k = 0, 1, \ldots, \overline{r}.$$

To complete the proof, it suffices to compose the function of form (30) and apply Theorem 2.

c) l is nonintegral and r is an integer.

Taking the cases analyzed above into consideration, we may assume without loss of generality that $m = n - 1$.

Let $l = \overline{l} + \lambda$, $\lambda > 0$, and let δ be small enough so that

$$\lambda - \frac{\delta (n-1)}{p^2} > 0.$$

We further choose an integer ρ so that

$$\rho-\left[l-\frac{n-1}{p}\right]>0,$$

$$\frac{n}{\rho-\left[l-\frac{n-1}{p}\right]}<\min\{p,\ p'-p,\ \delta\}\quad (p'>p). \tag{46}$$

We can then find an integer k so that

$$\frac{k-2}{\rho-\left[l-\frac{n-1}{p}\right]}<p,\qquad \frac{k}{\rho-\left[l-\frac{n-1}{p}\right]}>p.$$

It follows from (46) that $k>n$ and

$$p_1=\frac{k}{\rho-\left[l-\frac{n-1}{p}\right]}<p',\ p_1-p<p'-p.$$

By (46) we have

$$\lambda-(n-1)\left(\frac{1}{p}-\frac{1}{p_1}\right)=\lambda-(n-1)\frac{p_1-p}{pp_1}>\lambda-(n-1)\frac{p_1-p}{p^2}$$
$$>\lambda-\frac{(n-1)}{p^2}\left(\frac{2}{\rho-\left[l-\frac{n-1}{p}\right]}\right)\geqslant\lambda-\frac{n-1}{p^2}\delta>0.$$

Let $f\in W_p^l(R^{n-1})$. Since $r_1=l-(n-1)(1/p-1/p_1)$ is nonintegral, we have $f\in W_{p_1}^{r_1}(R^{n-1})$ and

$$\|f\|_{W_{p_1}^{r_1}(R^{n-1})}\leqslant c\|f\|_{W_p^l(R^{n-1})},$$

where c does not depend on f.

Then there exists a function $F\in W_{p_1}^{\rho}(R^k)$ such that

$$F|_{R^{n-1}}=f$$

and

$$\|F\|_{W_{p_1}^{\rho}(R^k)}\leqslant c\|f\|_{W_{p_1}^{r_1}(R^{n-1})}\leqslant c\|f\|_{W_p^r(R^{n-1})}. \tag{47}$$

In fact, it follows from Theorem 8 that there exists such an $F\in W_{p_1}^{\rho}(R^k)$, where

$$\bar{\rho} = r_1 + \frac{(k-n+1)}{p_1} = l - \frac{(n-1)}{p} + \frac{(n-1)}{p_1} + \frac{(k-n+1)}{p_1}$$

$$= l - \frac{n-1}{p} + \frac{k}{p_1} = l - \frac{(n-1)}{p} + \frac{k\left(\rho - \left[l - \frac{n-1}{p}\right]\right)}{k} = \rho.$$

To prove the theorem in our case, it suffices to apply Il′in's limit index theorem [24] to the function F. It is easily seen that a), b) and c) exhaust all possible cases.

The theorem is completely proved.

Corollary 1. *Let* $1 \le m \le n-1$, $1 < p < p' \le 2$ *and*

$$r = l + \frac{m-n-1}{p} - \left(\frac{1}{p} - \frac{1}{p'}\right) m > 0.$$

Then $W_p^l(R^{n-1}) \subset B_{p'}^r(R^m)$.

By Lemma 3, for integral l and r this result strengthens theorems of Sobolev [25] and Il′in [24].

Proof. It follows from the theorem that we can take l to be nonintegral, and $m = n-1$. Let $f \in W_p^l(R^{n-1})$; it then follows from (40) that there exists a function $F \in W^{\nu}_{p', p'(\nu-r)-1}(\bar{R}_0^n)$ such that $F|_{R^{n-1}} = f$ and

$$\| F \|_{W^{\nu}_{p', p'(\nu-r)-1}} \leqslant c \| f \|_{W_p^l(R^{n-1})},$$

where c does not depend on f. To complete the proof, it suffices to apply Corollary 2 of Theorem 7.

Corollary 2. *Let* $0 \le \alpha \le pl$, $1 \le m < n$, $1 < p < p' < \infty$ *and*

$$r = l - \frac{\alpha}{p} - \frac{(n-m)}{p} - m\left(\frac{1}{p} - \frac{1}{p'}\right) \gtrless 0.$$

Then

$$W_{p,\alpha}^l(\bar{R}_0^n) \subset W_{p'}^r(R_0^m).$$

This follows from Theorems 4 and 9.

Corollary 2 strengthens results of Kudrjavcev [22].

BIBLIOGRAPHY

[1] S. L. Sobolev, *Applications of functional analysis in mathematical physics*, Izdat. Leningrad. Gos. Univ., Leningrad, 1950; English transl., Transl. Math. Monographs, vol. 7, Amer. Math. Soc., Providence, R. I., 1963. MR 14, 565; 29 #2624.

[2] L. N. Slobodeckiĭ, *S. L. Sobolev's spaces of fractional order and their application to boundary problems for partial differential equations*, Dokl. Akad. Nauk SSSR **118** (1958), 243–246. (Russian) MR 21 #5059.

[3] ———, *Estimates of solutions of elliptic and parabolic systems*, Dokl. Akad. Nauk SSSR 120 (1958), 468–471. (Russian) MR 21 #5060.

[4] N. Aronszajn, *Boundary values of functions with finite Dirichlet integrals*, Conference on Partial Differential Equations, University of Kansas, 1955.

[5] L. N. Slobodeckiĭ and V. M. Babič, *On boundedness of the Dirichlet integrals*, Dokl. Akad. Nauk SSSR 106 (1955), 604–606. (Russian) MR 17, 959.

[6] L. N. Slobodeckiĭ, *Generalized Sobolev spaces and their application to boundary problems for partial differential equations*, Leningrad. Gos. Ped. Inst. Učen. Zap. 197 (1958), 54–112; English transl., Amer. Math. Soc. Transl. (2) 57 (1966), 207–276. MR **34** #3075.

[7] E. Gagliardo, *Caratterizzazioni delle tracce sulla frontiera relative ad alcune classi di funzioni in n variabili*, Rend. Sem. Math. Univ. Padova 27 (1957), 284–305. MR 21 #1525.

[8] L. N. Slobodeckiĭ, *Estimates in L_p of solutions of elliptic systems*, Dokl. Akad. Nauk SSSR 123 (1958), 616–619. (Russian) MR 21 #5061.

[9] O. V. Besov, *On some families of functional spaces. Imbedding and extension theorems*, Dokl. Akad. Nauk SSSR 126 (1959), 1163–1165. (Russian) MR 21 #5890.

[10] S. V. Uspenskiĭ, *An imbedding theorem for S.L. Sobolev's classes of fractional order*, Dokl. Akad. Nauk SSSR 130 (1960), 992–993 = Soviet Math. Dokl. 1 (1960), 132–133. MR 23 #A2042a.

[11] ———, *Properties of W_p^r classes with a fractional derivative on differentiable manifolds*, Dokl. Akad. Nauk SSSR 132 (1960), 60–62 = Soviet Math. Dokl. **1** (1960), 495–497. MR 23 #A2042b.

[12] ———, *Imbedding theorems for weight classes,* Trudy Mat. Inst. Steklov. 60 (1961), 282–303; English transl., Amer. Math. Soc. Transl. (2) 87 (1969), 121–145. MR 25 #440.

[13] J. E. Littlewood and R. E. A. C. Paley, *Theorems on Fourier series and power series.* II, Proc. London Math. Soc. (2) 42 (1936), 52–89.

[14] S. L. Sobolev, *Equations of mathematical physics,* OGIZ, Moscow, 1947; 3rd ed., 1954; English transl. of 3rd ed., Pergamon Press, London and Addison-Wesley, Reading, Mass., 1964. MR 16, 1027; 31 #2478.

[15] S. M. Nikol'skiĭ, *On the solution of the polyharmonic equation by a variational method,* Dokl. Akad. Nauk SSSR 88 (1953), 409–411. (Russian) MR 15, 425.

[16] V. M. Babič, *On the extension of functions,* Uspehi Mat. Nauk 8 (1953), no. 2 (54), 111–113. (Russian) MR 15, 110.

[17] G. M. Fihtengol'c, *A course of differential and integral calculus.* Vol. I, Fizmatgiz, Moscow, 1958; German transl., VEB Deutscher Verlag, Berlin, 1964.

[18] J. Marcinkiewicz, *Sur quelques intégrales du type de Dini,* Ann. Soc. Polon. Math. 17 (1938), 42–50.

[19] L. N. Slobodeckiĭ, *Imbedding of spaces* $W_p^{(l_1,\cdots,l_n)}$ *into Nikol'skiĭ* $H_p^{(l_1,\cdots,l_n)}$ *spaces,* Uspehi Mat. Nauk 15 (1960), no. 3 (93), 177–180. (Russian) MR 26 #5412.

[20] S. M. Nikol'skiĭ, *Inequalities for entire functions of exponential type and their application to the theory of differentiable functions of several variables,* Trudy Mat. Inst. Steklov. 38 (1951), 244–278; English transl., Amer. Math. Soc. Transl. (2) 80 (1969), 1–38. MR 14, 32

[21] ———, *Properties of certain classes of functions of several variables on differentiable manifolds,* Mat. Sb. 33 (75) (1953), 261–326; English transl., Amer. Math. Soc. Transl. (2) 80 (1969), 39–118. MR 16, 453.

[22] L. D. Kudrjavcev, *Direct and inverse imbedding theorems. Applications to the solution of elliptic equations by variational methods,* Trudy Mat. Inst. Steklov. 55 (1959). (Russian) MR 33 #7838.

[23] P. I. Lizorkin, *Boundary properties of functions from "weight" classes,* Dokl. Akad. Nauk SSSR 132 (1960), 514–517 = Soviet Math. Dokl. 1 (1960), 589–593. MR 23 #A434.

[24] V. P. Il′in, *On a theorem of inclusion for a limiting exponent*, Dokl. Akad. Nauk SSSR 96 (1954), 905–908. (Russian) MR 16, 121.

[25] S. Sobolev, *On a theorem of functional analysis*, Mat. Sb. **4** (**46**) (1938), 471–497; English transl., Amer. Math. Soc. Transl. (2) **34** (1963), 39–68.

[26] V. P. Il′in and V. A. Solonnikov, *Some properties of differentiable functions of several variables*, Dokl. Akad. Nauk SSSR **136** (1961), 538–541 = Soviet Math. Dokl. 2 (1961), 95–98. MR 22 #8258.

Translated by:
N. Koblitz

Amer. Math. Soc. Transl.
(2) Vol. 90, 1970

ON FACTORIZATION OF OPERATORS RELATIVE TO A DISCRETE CHAIN OF PROJECTORS IN BANACH SPACE*

M. A. BARKAR' AND I. C. GOHBERG

The abstract problem of factorization of operators relative to a chain of projectors was posed and solved for a wide class of operators operating in Hilbert space in the paper [1] of M. G. Kreĭn and one of the authors.

In the present paper we generalize some of these results to operators which operate in Banach space for the case of a discrete chain. The case of a continuous chain will be considered in another publication.

The paper consists of three sections. In the first we consider the problem of factorization of operators relative to a finite chain. In the second we establish general theorems on the factorization of operators relative to a discrete chain, and in the third we distinguish concrete classes of operators admitting factorization.

The authors express their gratitude to A. S. Markus for useful discussions of the results of this paper.

§ 1. Factorization of operators relative to a finite chain

1. We denote by $\mathfrak{R}(\mathfrak{B})$ the ring of all bounded linear operators operating in some Banach space $\mathfrak{B}$.

In the set of all projectors of $\mathfrak{R}(\mathfrak{B})$ we may introduce a partial ordering, putting $P_1 < P_2$ if $P_1\mathfrak{B} \subset P_2\mathfrak{B}$ and $(I - P_1)\mathfrak{B} \supset (I - P_2)\mathfrak{B}$.

Every ordered set of projectors containing the projectors 0 and I is said to be a *chain*. If $\mathfrak{P} = \{P\}$ is a chain, then the set of all projectors $\mathfrak{Q} = \{Q\}$, $Q = I - P$, $P \in \mathfrak{P}$ is also a chain. The chain $\mathfrak{Q}$ will be said to be *complementary* to the chain $\mathfrak{P}$. If $\mathfrak{P}$ contains a finite number of projectors, it is said to be finite.

The chain $\mathfrak{P}$ is said to be a *proper* chain of the operator $A \in \mathfrak{R}(\mathfrak{B})$ if for any operator $P \in \mathfrak{P}$ the subspace $P\mathfrak{B}$ is invariant relative to the operator A, i.e. $PAP = AP$.

The set of operators of $\mathfrak{R}(\mathfrak{B})$ which have the chain $\mathfrak{P}$ as a proper chain

---

*) Translation of Mat. Issled. 1 (1966), no. 1, 32–54.

will be denoted by $\mathfrak{R}_+(\mathfrak{P})$. Analogously the set of operators of $\mathfrak{R}(\mathfrak{B})$ having the complementary chain $\mathfrak{Q}$ as a proper chain will be denoted by $\mathfrak{R}_-(\mathfrak{P})$. Finally by $\mathfrak{R}_0(\mathfrak{P})$ we denote the set of operators of $\mathfrak{R}(\mathfrak{B})$ having both chains $\mathfrak{P}$ and $\mathfrak{Q}$ as proper chains. Obviously $\mathfrak{R}_\pm(\mathfrak{P}) = \mathfrak{R}_\mp(\mathfrak{Q})$ and $\mathfrak{R}_0(\mathfrak{P}) = \mathfrak{R}_0(\mathfrak{Q})$.

It is easy to prove the following assertion.

1°. *For the operator X of $\mathfrak{R}_\pm(\mathfrak{P})$ to be invertible and for the operator X^{-1} to lie in $\mathfrak{R}_\pm(\mathfrak{P})$, it is necessary and sufficient that the operator $\Sigma_{j=1}^n \Delta P_j X \Delta P_j$ $(\Delta P_j = P_j - P_{j-1})$ be invertible.*

We shall say that the invertible operator A admits a factorization relative to the finite chain $\mathfrak{P} = \{0 = P_0 < P_1 < \cdots < P_n = I\}$ if it is representable in the form

$$A = X_- S X_+, \tag{1.1}$$

where $X_\pm \in \mathfrak{R}_\pm(\mathfrak{P})$, $D \in \mathfrak{R}_0(\mathfrak{P})$ and $\Delta P_j\ X_\pm \Delta P_j = \Delta P_j$ $(j = 1, 2, \cdots, n)$.

Since the operator $\Sigma_{j=1}^n \Delta P_j X_\pm \Delta P_j = I$, it follows that the operator $X_\pm$ is invertible, and along with it the operator D as well, and $X_\mp^{-1} \in \mathfrak{R}_\pm(\mathfrak{P})$, while

$$\Delta P_j X_\pm^{-1} \Delta P_j = \Delta P_j \quad (j = 1, 2, \cdots, n).$$

We shall show that if an invertible operator A admits a factorization (1.1), then the factors $X_\pm$ and D are determined uniquely.

Indeed, suppose that $A = X_- D X_+$ and $A = Y_- D' Y_+$, while

$$X_\pm, Y_\pm \in \mathfrak{R}_\pm(\mathfrak{P});\ D_1 D' \in \mathfrak{R}_0(\mathfrak{P}) \tag{1.2}$$

and

$$\Delta P_j X_\pm \Delta P_j = \Delta P_j,\ \Delta P_j Y_\pm \Delta P_j = \Delta P_j \quad (j = 1, 2, \cdots, n). \tag{1.3}$$

Then

$$Y_-^{-1} X_- = D' Y_+ X_+^{-1} D. \tag{1.4}$$

It follows from (1.2) and (1.4) that

$$Y_-^{-1} X_- = \sum_{j=1}^n \Delta P_j Y_-^{-1} X_- \Delta P_j$$

or

$$Y_-^{-1}X_- = \sum_{j=1}^{n} \Delta P_j Y_-^{-1} \Delta P_j \sum_{j=1}^{n} \Delta P_j X_- \Delta P_j$$

and therefore, in view of (1.3), $Y_-^{-1}X_- = I$, that is $Y_- = X_-$.

Analogously one verifies that $Y_+ = X_+$, so that then $D' = D$ as well.

Theorem 1. *For the invertible operator A to have a factorization* (1.1) *relative to the chain $\mathfrak{P}\{0 = P_0 < P_1 < \cdots < P_n = I\}$, it is necessary and sufficient that the operator P_jAP_j $(j = 1, 2, \cdots, n)$ should be invertible in the corresponding subspaces $P_j\mathfrak{B}$. If this condition is satisfied, then the following equations hold for the factors of the factorization:*[1]

$$X_- = \overset{\leftarrow}{\prod_{j=1}^{n}} (I + \Delta P_j A P_{j-1} (P_{j-1} A P_{j-1})^{-1}), \tag{1.5}$$

$$X_+ = \prod_{j=1}^{n} (I + (P_{j-1} A P_{j-1})^{-1} P_{j-1} A \Delta P_j), \tag{1.6}$$

$$D = \sum_{j=1}^{n} (\Delta P_j A \Delta P_j - \Delta P_j A P_{j-1} (P_{j-1} A P_{j-1})^{-1} P_{j-1} A \Delta P_j), \tag{1.7}$$

and for the inverse operators, the equations[2]

$$X_-^{-1} = I - \sum_{j-1}^{n} \Delta P_j A P_{j-1} (P_{j-1} A P_{j-1})^{-1}, \tag{1.8}$$

$$X_+^{-1} = I - \sum_{j=1}^{n} (P_{j-1} A P_{j-1})^{-1} P_{j-1} A \Delta P_j, \tag{1.9}$$

$$D^{-1} = \sum_{j=1}^{n} \Delta P_j (P_j A P_j)^{-1} \Delta P_j. \tag{1.10}$$

Proof. If the invertible operator A admits the factorization (1.1), then

$$P_j A P_j = (P_j X_- P_j) D (P_j X_+ P_j) \quad (j = 1, 2, \cdots, n).$$

---

1) We denote the product $A_n A_{n-1} \cdots A_1$ by $\overset{\leftarrow}{\prod_{j=1}^{n}} A_j$.

2) These equations are essentially contained in [1].

Since in the subspace $P_j\mathfrak{B}$ the operators $P_jX_-P_j$ and $P_jX_+P_j$ are invertible,[1] and $(P_jX_-P_j)^{-1} = P_jX_-^{-1}P_j$ and $(P_jX_+P_j)^{-1} = P_jX_+^{-1}P_j$, in this subspace the operator P_jAP_j is invertible as well and

$$(P_jAP_j)^{-1} = (P_jX_+^{-1}P_j)D^{-1}(P_jX_-^{-1}P_j) \quad (j = 1, 2, \cdots, n).$$

Now suppose that the operators P_jAP_j $(j = 1, 2, \cdots, n)$ are invertible in the corresponding subspaces $P_j\mathfrak{B}$.

By a direct calculation one then verifies the equation

$$P_jAP_j = \Lambda_j^-(P_{j-1}AP_{j-1} + D_j)\Lambda_j^+, \quad (j = 1, 2, \cdots, n) \tag{1.11}$$

where

$$\Lambda_j^- = I + \Delta P_jAP_{j-1}\,(P_{j-1}AP_{j-1})^{-1}, \tag{1.12}$$

$$\Lambda_j^+ = I + (P_{j-1}AP_{j-1})^{-1}P_{j-1}A\Delta P_j, \tag{1.13}$$

$$D_j = \Delta P_jA\Delta P_j - \Delta P_jAP_{j-1}\,(P_{j-1}AP_{j-1})^{-1}P_{j-1}A\Delta P_j. \tag{1.14}$$

Noting that

$$\Lambda_k^- D_j \Lambda_k^+ = D_j \quad (k < j = 2, \cdots, n),$$

we find from (1.11) that

$$A = \overleftarrow{\prod_{j=1}^{n}}\Lambda_j^-\left\{\sum_{j=1}^{n} D_j\right\}\prod_{j=1}^{n}\Lambda_j^+, \tag{1.15}$$

where Λ_j and D_j $(j = 1, 2, \cdots, n)$ are defined by the formulas (1.12)–(1.14).

It is easy to verify that the operators

$$X_- = \overleftarrow{\prod_{j=1}^{n}}\Lambda_j^-, \quad X_+ = \prod_{j=1}^{n}\Lambda_j^+, \quad D = \sum_{j=1}^{n} D_j$$

---

1) In what follows we denote by $(P_jAP_j)^{-1}$ the operator inverse to the operator P_jAP_j in the subspace $P_j\mathfrak{B}$ and equal to zero on the subspace $(I - P_j)\mathfrak{B}$.

lie respectively in the sets $\mathfrak{R}_-(\mathfrak{P})$, $\mathfrak{R}_+(\mathfrak{P})$, $\mathfrak{R}_0(\mathfrak{P})$ and that $\Delta P_j X_\pm \Delta P_j = \Delta P_j$ ($j = 1, 2, \cdots, n$). Accordingly, equation (1.15) yields a factorization (1.1) of A relative to the chain $\mathfrak{P} = \{0 = P_0 < P_1 < \cdots < P_n = I\}$, and for the factors $X_\pm$ and D of this factorization the formulas (1.5)–(1.7) result.

Formulas (1.8) and (1.9) follow immediately from the equations

$$(I + \Delta P_j A P_{j-1}(P_{j-1}AP_{j-1})^{-1})^{-1} = I - \Delta P_j A P_{j-1}(P_{j-1}AP_{j-1})^{-1},$$

$$P_{j-1}\Delta P_k = 0, \; j \le k$$

and

$$(I + (P_{j-1}AP_{j-1})^{-1}P_{j-1}A\Delta P_j)^{-1} = I - (P_{j-1}AP_{j-1})^{-1}P_{j-1}A\Delta P_j,$$

$$\Delta P_j P_{k-1} = 0, \; j \ge k,$$

Finally, in order to obtain formula (1.10) we represent the operator A in the form $A = X_- \widetilde{X}_+$ ($\widetilde{X}_+ = DX_+$). Then

$$(P_jAP_j)^{-1}\Delta P_j = \widetilde{X}_+^{-1} P_j X_-^{-1}\Delta P_j = \widetilde{X}_+^{-1}(\Delta P_j + P_{j-1})X_-^{-1}\Delta P_j,$$

or

$$(P_jAP_j)^{-1}\Delta P_j = \widetilde{X}_+^{-1}\Delta P_j \quad (j = 1, 2, \cdots, n).$$

Hence

$$\widetilde{X}_+^{-1} = X_+^{-1}D^{-1} = \sum_{j=1}^{n} (P_jAP_j)^{-1}\Delta P_j$$

and

$$\Delta P_j X_+^{-1} D^{-1}\Delta P_j = \Delta P_j D^{-1} = \Delta P_j (P_jAP_j)^{-1}\Delta P_j,$$

so that

$$D^{-1} = \sum_{j=1}^{n} \Delta P_j (P_jAP_j)^{-1}\Delta P_j.$$

The theorem is proved.

We note that in the proof of the theorem we have along the way found the formula

$$X_{+}^{-1}D^{-1} = \sum_{j=1}^{n} (P_jAP_j)^{-1}\Delta P_j. \tag{1.16}$$

Analogously we may also obtain the formula

$$D^{-1}X_{-}^{-1} = \sum_{j=1}^{n} \Delta P_j(P_jAP_j)^{-1}. \tag{1.17}$$

Theorem 1, for the case $n = 2$, was proved in a more awkward way in the paper [2], where it was assumed that A is an invertible positive operator operating in Hilbert space. Under this hypothesis the condition of invertibility of the operators P_jAP_j is fulfilled automatically. We note that the same situation holds for any invertible dissipative operator $((A - A^*)/2i \geq 0)$ operating in Hilbert space.

2. Suppose that the invertible operator A admits a factorization (1.1) relative to the chain $\mathfrak{P} = \{0 = P_0 < P_1 < \cdots < P_n = I\}$. Then

$$A^{-1} = X_{+}^{-1}D^{-1}X_{-}^{-1}. \tag{1.18}$$

Consider the complementary chain

$$\mathfrak{Q} = \{0 = Q_0 < Q_1 < \cdots < Q_n = I\},\ Q_j = I - P_{n-j}\ (j = 1, 2, \cdots, n).$$

Since $X_{\pm}^{-1} \in \mathfrak{R}_{\pm}(\mathfrak{P})$, we have $X_{\pm}^{-1} \in \mathfrak{R}_{\mp}(\mathfrak{Q})$ and $D \in \mathfrak{R}_0(\mathfrak{Q})$. Moreover, $\Delta Q_j X_{\pm}^{-1}\Delta Q_j = \Delta Q_j$ $(j = 1, 2, \cdots, n)$. Accordingly, equation (1.18) yields a factorization of the operator A^{-1} relative to the chain $\mathfrak{Q}$.

Thus, if the operators P_jAP_j are invertible in the subspaces $P_j\mathfrak{B}$, then the operators $Q_jA^{-1}Q_j$ are invertible in the subspaces $Q_j\mathfrak{B}$, and the following equations hold:

$$\overleftarrow{\prod_{j=1}^{n}} (I + \Delta P_jAP_{j-1}(P_{j-1}AP_{j-1})^{-1}) = I - \sum_{j=1}^{n} (Q_{j-1}A^{-1}Q_{j-1})^{-1}Q_{j-1}A^{-1}\Delta Q_j, \tag{1.19}$$

$$\prod_{j=1}^{n} (I + (P_{j-1}AP_{j-1})^{-1}P_{j-1}A\Delta P_j) = I - \sum_{j=1}^{n} \Delta Q_jA^{-1}Q_{j-1}(Q_{j-1}A^{-1}Q_{j-1})^{-1}, \tag{1.20}$$

$$\prod_{j=1}^{n} (\Delta P_jA\Delta P_j - \Delta P_jAP_{j-1}(P_{j-1}AP_{j-1})^{-1}P_{j-1}A\Delta P_j) = \sum_{j=1}^{n} \Delta Q_j(Q_jA^{-1}Q_j)^{-1}\Delta Q_j, \tag{1.21}$$

and also the equations

$$\sum_{j=1}^{n}(P_jAP_j)^{-1}\Delta P_j = \left[\sum_{j=1}^{n}\Delta Q_j(Q_jA^{-1}Q_j)^{-1}\right]^{-1}, \tag{1.22}$$

$$\sum_{j=1}^{n}\Delta P_j(P_jAP_j)^{-1} = \left[\sum_{j=1}^{n}(Q_jA^{-1}Q_j)^{-1}\Delta Q_j\right]^{-1}. \tag{1.23}$$

The propositions presented in this section have a purely algebraic character. They may be formulated for any ring with unity.

§ 2. General theorems on the factorization of operators relative to a discrete chain

1. Consider in Banach space $\mathfrak{B}$ the chain of projectors

$$\mathfrak{P} = \{0 = P_0 < P_1 < \cdots < P_j < \cdots < I\}, \tag{2.1}$$

where $\dim \Delta P_j\mathfrak{B} = 1$, $\Delta P_j = P_j - P_{j-1}$ $(j = 1, 2, \cdots)$ and P_j converges strongly to I as $j \to \infty$. We shall call such a chain *discrete*.

We shall say that the invertible operator A admits factorization relative to the chain (2.1) if it is representable in the form

$$A = X_-DX_+, \tag{2.2}$$

where $X_\pm$ are invertible operators from $\mathfrak{R}_\pm(\mathfrak{P})$, and $D \in \mathfrak{R}_0(\mathfrak{P})$, and

$$\Delta P_jX_\pm\Delta P_j = \Delta P_j \quad (j = 1, 2, \cdots).$$

It is easy to see that if the invertible operator A admits factorization, then $X_\pm^{-1} \in \mathfrak{R}_\pm(\mathfrak{P})$ and $D^{-1} \in \mathfrak{R}_0(\mathfrak{P})$.

Just as in the case of a finite chain, one proves that if the invertible operator A admits a factorization (2.2) relative to the chain (2.1), then the factors $X_\pm$ and D are uniquely defined.

In what follows we restrict ourselves to chains $\mathfrak{P}$ which have the following property: (α) *for any* $x \in \mathfrak{B}$ *the series* $\sum_{j=1}^{\infty}\Delta P_jx$ *converges unconditionally*. Accordingly, if one takes a unit vector e_j in each of the subspaces $\Delta P_j\mathfrak{B}$ $(j = 1, 2, \cdots)$, the system $\{e_j\}_1^\infty$ forms an unconditional basis of the space $\mathfrak{B}$.

In this case, as is known (see [3], Theorem 7), there exists a constant M such that for any numbers $\{\beta_j\}$ and $\{\alpha_j\}$ the following relation holds:

$$\left|\sum_j \beta_j\alpha_je_j\right| \leq M \sup|\beta_j| \left|\sum_j \alpha_je_j\right|. \tag{2.3}$$

Theorem 2. *If the invertible operator* A *admits a factorization* (2.2) *relative to the chain* (2.1), *then:*

a) *The operators* P_jAP_j $(j = 1, 2, \cdots)$ *are invertible in the subspaces* $P_j\mathfrak{B}$, *and*

$$\sup_j |(P_jAP_j)^{-1}| < \infty. \tag{2.4}$$

b) *For the factors* $X_\pm$ *and* D *the equations*

$$X_- = \overleftarrow{\prod_{j=1}^{\infty}}(I + \Delta P_jAP_{j-1}(P_{j-1}AP_{j-1})^{-1}), \tag{2.5}$$

$$X_+ = \prod_{j=1}^{\infty}(I + (P_{j-1}AP_{j-1})^{-1}P_{j-1}A\Delta P_j), \tag{2.6}$$

$$D = \sum_{j=1}^{\infty}(\Delta P_jA\Delta P_j - \Delta P_jAP_{j-1}(P_{j-1}AP_{j-1})^{-1}P_{j-1}A\Delta P_j \tag{2.7}$$

hold, and for their inverses the equations

$$X_-^{-1} = I - \sum_{j=1}^{\infty}\Delta P_jAP_{j-1}(P_{j-1}AP_{j-1})^{-1}, \tag{2.8}$$

$$X_+^{-1} = I - \sum_{j=1}^{\infty}(P_{j-1}AP_{j-1})^{-1}P_{j-1}A\Delta P_j, \tag{2.9}$$

$$D^{-1} = \sum_{j=1}^{\infty}\Delta P_j(P_jAP_j)^{-1}\Delta P_j, \tag{2.10}$$

while all the series and products in (2.5)–(2.10) *converge unconditionally.*

Conversely, if condition a) *is satisfied and if at least one of the following pairs of series and products converges:* (2.5) *and* (2.6); (2.8) *and* (2.9); (2.5) *and* (2.8); (2.6) *and* (2.9), *then the operator* A *admits factorization relative to the chain* (2.1).

We preface the proof of the theorem with the following proposition.

Lemma 1. *If condition* a) *is satisfied for the invertible operator* A, *then the series* (2.7) *and* (2.10) *strongly converge to mutually inverse operators.*

Proof. We denote the partial sums of the series (2.7) and (2.10) respectively by D_n and $\widetilde{D}_n$.

Then for any vector $x = \sum_{j=1}^{\infty}\alpha_je_j$ from $\mathfrak{B}$

$$|\widetilde{D}_n x| = \left|\sum_{j=1}^{n} \Delta P_j (P_j A P_j)^{-1} \Delta P_j \alpha_j e_j\right| = \left|\sum_{j=1}^{n} \beta_j \alpha_j e_j\right|, \tag{2.11}$$

where $\beta_j e_j = \Delta P_j (P_j A P_j)^{-1} \Delta P_j e_j$.

In view of (2.3)

$$|\widetilde{D}_n x| \leq M \sup_j |\beta_j| \left|\sum_{j=1}^{n} \alpha_j e_j\right|.$$

Since $|\beta_j| = |\Delta P_j (P_j A P_j)^{-1} \Delta P_j e_j| \leq |\Delta P_j (P_j A P_j)^{-1} \Delta P_j|$, and the sequence $\{P_j\}_1^\infty$ is bounded and condition a) is satisfied, it follows that $\sup_n |\widetilde{D}_n| < \infty$. Analogously one establishes that $\sup_n |D_n| < \infty$. Moreover, the sequences $\{D_n\}_1^\infty$ and $\{\widetilde{D}_n\}_1^\infty$ converge on the set $\bigcup_j P_j \mathfrak{B}$, dense in $\mathfrak{B}$, so that they converge everywhere.

It follows from the equation $D_n \widetilde{D}_n = P_n$ that the sequences $\{D_n\}_1^\infty$ and $\{\widetilde{D}_n\}_1^\infty$ converge and are mutually inverse operators. The lemma is proved.

It is easy to see that equation (2.7) may be rewritten in the following form:

$$D = \sum_{j=1}^{\infty} \frac{d_j}{d_{j-1}} \Delta P_j,$$

where $d_j = \det(P_j A P_j)$ $(j = 1, 2, \cdots)$, $d_0 = 1$.

Then obviously D^{-1} has the form

$$D^{-1} = \sum_{j=1}^{\infty} \frac{d_{j-1}}{d_j} \Delta P_j.$$

Proof of Theorem 2. Suppose that the operator A admits the factorization (2.2) relative to the chain (2.1).

The invertibility of the operators $P_j A P_j$ $(j = 1, 2, \cdots)$ in the subspaces $P_j \mathfrak{B}$ follows from the equation

$$P_j A P_j = (P_j X_- P_j) D (P_j X_+ P_j). \tag{2.12}$$

Hence we find that $\sup_j |(P_j A P_j)^{-1}| < \infty$.

Equation (2.12) may also be written in the form

$$P_n A P_n = (P_n X_- P_n)(P_n D P_n)(P_n X_+ P_n). \tag{2.13}$$

It yields a factorization of the operator $P_n A P_n$ in the subspace $P_n \mathfrak{B}$ relative to the chain $\mathfrak{P}_n = \{0 = P_0 < P_1 < \cdots < P_n\}$.

Accordingly, in view of Theorem 1,

$$P_nX_-P_n = P_n\overleftarrow{\prod_{j=1}^{n}}(I + \Delta P_jAP_{j-1}(P_{j-1}AP_{j-1})^{-1}) \tag{2.14}$$

$$P_nX_+P_n = P_n\prod_{j=1}^{n}(I + (P_{j-1}AP_{j-1})^{-1}P_{j-1}A\Delta P_j), \tag{2.15}$$

$$P_nDP_n = \sum_{j=1}^{n}(\Delta P_jA\Delta P_j - \Delta P_jAP_{j-1}(P_{j-1}AP_{j-1})^{-1}P_{j-1}A\Delta P_j). \tag{2.16}$$

Equations (2.5), (2.6) and (2.7), as well as the strong convergence on their right sides, follow from the above equations.

Rewrite equation (2.13) in the form

$$P_nX_-^{-1}P_n = (P_nDP_n)(P_nX_+P_n)(P_nAP_n)^{-1}.$$

Taking account of Theorem 1, we arrive at formula (2.8), its right side converging strongly. We prove formulas (2.9) and (2.10) analogously.

Suppose now that condition a) is satisfied and that the products (2.5) and (2.6) converge. We denote their limits by X_- and X_+ respectively. Obviously $X_\pm \in \mathfrak{R}_\pm(\mathfrak{P})$ and $\Delta P_jX_\pm\Delta P_j = \Delta P_j$ $(j = 1, 2, \cdots)$.

Suppose that the equation

$$P_nAP_n = X_{n-}D_nX_{n+} \tag{2.17}$$

yields a factorization of the operator P_nAP_n relative to the chain $\mathfrak{P}_n = \{0 = P_0 < P_1 < \cdots < P_n\}$ in the subspace $P_n\mathfrak{B}$.

It is easy to see that X_{n-} and X_{n+} are equal respectively to the right sides of equations (2.14) and (2.15), and that D_n is the partial sum of the right side of equation (2.7). By Lemma 1 the sequence $\{D_n\}_1^\infty$ converges. We denote its limit by D. Obviously $D \in \mathfrak{R}_0(\mathfrak{P})$.

Since

$$X_n^{-1} = D_nX_{n+}(P_nAP_n)^{-1},$$

the limit of X_{n-}^{-1} exists as $n \longrightarrow \infty$. It is easy to see that it is equal to the operator X_-^{-1}.

Analogously one proves that the limit of X_{n+}^{-1} is X_+^{-1}.

Now we pass to the limit as $n \longrightarrow \infty$ in equation (2.17). The result is the equation

$$A = X_- D X_+,$$

which yields a factorization of the operator A relative to the chain $\mathfrak{P}$.

Now suppose that condition a) is satisfied, and that, for example the product (2.5) and the series (2.8) converge.

From the equation

$$Y_{n+} = D_n^{-1} X_n^{-1} (P_n A P_n),$$

and from Lemma 1 and the convergence of the series (2.8), it follows that the product (2.6) converges, and then, from what has already been proved, the operator A admits the factorization (2.2) relative to the chain (2.1)

Analogously one treats the other two cases. The theorem is proved.

Corollary 1. *If at least one of the products* (2.5), (2.6) *or at least one of the series* (2.8), (2.9) *converges uniformly, then the operator A admits the factorization* (2.2) *relative to the chain* (2.1).

Indeed, suppose for example that the series (2.8) converges uniformly. Then, inasmuch as the nth power of the operator $\Sigma_{j=1}^{n} \Delta P_j A P_{j-1} (P_{j-1} A P_{j-1})$ is equal to zero, that operator is a Volterra operator. But, as is known [4], the uniform limit of a sequence of Volterra operators is also a Volterra operator.[1)] Accordingly the sum of the series (2.8) is an invertible operator. In such a case the product (2.5) converges, also uniformly, and because of Theorem 2, the operator A admits a factorization (2.2) relative to the chain (2.1).

2. In the present subsection we present some complements to Theorem 2 for the case when the operator $A - I = T$ is completely continuous. If the operator $A - I$ is completely continuous, equations (2.7) and (2.10) may be rewritten in the form

$$D = I + \sum_{j=1}^{\infty} (\Delta P_j T \Delta P_j - \Delta P_j T P_{j-1} (I + P_{j-1} T P_{j-1})^{-1} P_{j-1} T \Delta P_j). \qquad (2.7')$$

$$D^{-1} = I - \sum_{j=1}^{\infty} \Delta P_j T P_j (I + P_j T P_j)^{-1} \Delta P_j. \qquad (2.10')$$

1) In [4] (Chapter I, §4) this proposition was proved for a Hilbert space. However the proof given there remains valid for a Banach space as well.

As a complement to Lemma 1 we have the following proposition.

Lemma 2. *If the operator $A - I = T$ is completely continuous, then the series (2.7') and (2.10') converge uniformly to mutually inverse operators D and D'. The operators $D - I$ and $D' - I$ are completely continuous.*

Proof. For any vector $x = \sum_{j=1}^{\infty} \alpha_j e_j$ of $\mathfrak{B}$

$$\left| \sum_{j=n}^{\infty} \Delta P_j T P_j (I + P_j T P_j)^{-1} \Delta P_j x \right|$$

$$\leq M \sup_{j \geq n} |\Delta P_j T P_j (I + P_j T P_j)^{-1} \Delta P_j| |x| \leq M_1 \sup_{j \geq n} |\Delta P_j T| |x|. \tag{2.18}$$

The sequence $\{\Delta P_j\}_1^{\infty}$ converges strongly to zero as $j \to \infty$, and T is a completely continuous operator. Accordingly the product $\Delta P_j T$ converges uniformly to zero, and, in view of (2.18), the series (2.10') converges uniformly.

Analogously one proves the uniform convergence of the series (2.7'). The lemma is proved.

2°. *If the operator A admits a factorization and the operator $X_- - I$ is completely continuous, then the product (2.5) and the series (2.8) converge uniformly.*

Indeed, write equation (2.2) in the form

$$I + B = (I + T)(I + C),$$

where

$$I + B = A, \; I + T = X_-, \; I + C = DX_+.$$

We successively obtain

$$I + P_{j-1} B P_{j-1} = (I + P_{j-1} T P_{j-1})(I + P_{j-1} C P_{j-1}),$$

$$B P_{j-1} (I + P_{j-1} B P_{j-1})^{-1}$$

$$= [C P_{j-1} + T P_{j-1} (I + P_{j-1} C P_{j-1})](I + P_{j-1} C P_{j-1})^{-1} (I + P_{j-1} T P_{j-1})^{-1},$$

or, since $\Delta P_j C P_{j-1} = 0$,

$$\Delta P_j B P_{j-1} (I + P_{j-1} B P_{j-1})^{-1} = \Delta P_j T P_{j-1} (I + P_{j-1} T P_{j-1})^{-1},$$

or

$$\Delta P_j B P_{j-1}(I + P_{j-1}BP_{j-1})^{-1} = \Delta P_j T P_{j-1}(I + T)^{-1}.$$

But, since $P_{j-1} = P_j - \Delta P_j$ and $\Delta P_j T \Delta P_j = 0$, we have

$$\Delta P_j B P_{j-1}(I + P_{j-1}BP_{j-1})^{-1} = \Delta P_j T(I + T)^{-1}$$

and

$$\sum_{j=1}^{n} \Delta P_j B P_{j-1}(I + P_{j-1}BP_{j-1})^{-1} = P_n T(I + T)^{-1}.$$

It follows from this equation that the series (2.8) converges uniformly to the operator

$$I - T(I + T)^{-1} = (I + T)^{-1} = X_-^{-1}.$$

The uniform convergence of the series (2.8) implies the uniform convergence of the product (2.5).

We note that the analogous assertion for the operator X_+ is false.

It is easy to see that the following proposition holds.

3°. *If the operator A admits a factorization (2.2) relative to the chain (2.1) and the operators $A - I = T$ and $X_- - I = T_-$ are completely continuous, then for the series (2.9) and product (2.6) to be uniformly convergent it is necessary and sufficient that the sequence of operators $\{P_j T P_j\}_1^\infty$ converge uniformly to the operator T.*

The following is an easy consequence of Lemma 2 and propositions 2° and 3°.

Theorem 3. *Suppose that the operator $A - I = T$ is completely continuous, the operators $I + P_j T P_j$ invertible and the sequence $\{P_j T P_j\}_1^\infty$ converges uniformly to the operator T. For the operator A to admit a factorization (2.2) relative to the chain (2.1) and for the operators $X_\pm - I$ and $D - I$ to be completely continuous, ·· is necessary that the products (2.5), (2.6) and the series (2.7′), (2.8), (2.9) and (2.10′) converge uniformly, and sufficient, that at least one of the products (2.5), (2.6) or one of the series (2.8), (2.9) should converge uniformly.*

We note the following. 1) If the sequence $\{P_j^*\}$ converges strongly to the unit operator in the space $\mathfrak{B}$, or 2) if the space $\mathfrak{B}$ is reflexive, then the sequence $\{P_j T P_j\}_1^\infty$ converges uniformly to the operator T.

3. Suppose that

$$\mathfrak{Q} = \{I = Q_0 > Q_1 > \cdots > Q_j > \cdots ;\ 0\}$$

is the complementary chain to the discrete chain $\mathfrak{P}$. All of the propositions proved above on the factorization of an operator relative to a discrete chain $\mathfrak{P}$ remain in force for factorization relative to the chain $\mathfrak{Q}$.

Between the factorizations relative to the chains $\mathfrak{P}$ and $\mathfrak{Q}$ there exists the following connection. If the operator A admits a factorization (2.2) relative to the chain $\mathfrak{P}$, then the equation

$$A^{-1} = X_+^{-1} D^{-1} X_-^{-1}$$

yields a factorization of the operator A^{-1} relative to the chain $\mathfrak{Q}$, and the following equations obviously hold:

$$\overleftarrow{\prod_{j=1}^{\infty}} (I + \Delta P_j A P_{j-1} (P_{j-1} A P_{j-1})^{-1}) = I - \sum_{j=1}^{\infty} (Q_j A^{-1} Q_j)^{-1} Q_j A^{-1} \Delta Q_j, \quad (2.19)$$

$$\prod_{j=1}^{\infty} (I + (P_{j-1} A P_{j-1})^{-1} P_{j-1} A \Delta P_j) = I - \sum_{j=1}^{\infty} \Delta Q_j A^{-1} Q_j (Q_j A^{-1} Q_j)^{-1}, \quad (2.20)$$

$$\sum_{j=1}^{\infty} (\Delta P_j A \Delta P_j - \Delta P_j A P_{j-1} (P_{j-1} A P_{j-1})^{-1} P_{j-1} A \Delta P_j)$$

$$= \sum_{j=1}^{\infty} \Delta Q_j (Q_{j-1} A^{-1} Q_{j-1})^{-1} \Delta Q_j, \quad (2.21)$$

and also the equations

$$\sum_{j=1}^{\infty} (P_j A P_j)^{-1} \Delta P_j = \left[\sum_{j=1}^{\infty} \Delta Q_j (Q_{j-1} A^{-1} Q_{j-1})^{-1} \right]^{-1}, \quad (2.22)$$

$$\sum_{j=1}^{\infty} \Delta P_j (P_j A P_j)^{-1} = \left[\sum_{j=1}^{\infty} (Q_{j-1} A^{-1} Q_{j-1})^{-1} \Delta Q_j \right]^{-1}. \quad (2.23)$$

§ 3. Theorems on the factorization of operators close to unity

1. An operator T operating in a Banach space $\mathfrak{B}$ is said to be *nuclear* if it is representable in the form

$$Tx = \sum_{r=1}^{\infty} a_r \phi_r(x) \eta_r \quad (x \in \mathfrak{B}), \quad (3.1)$$

where the a_r are complex numbers satisfying the condition

$$\sum_{r=1}^{\infty} |a_r| < \infty,$$

and $\phi_r \in \mathfrak{B}^*$, $\eta_r \in \mathfrak{B}$ and $|\phi_r| = |\eta_r| = 1$ $(r = 1, 2, \cdots)$.

To each nuclear operator there corresponds the nuclear norm

$$\|T\| = \inf \sum_{r=1}^{\infty} |a_r|,$$

the infimum being taken over all representations of T in the form (3.1).

As in § 2, we will suppose that $\mathfrak{P}$ is a discrete chain satisfying the condition (α).

Lemma 3. *Suppose that* $\{G_j\}_1^\infty$ *is a bounded sequence of operators of* $\mathfrak{R}(\mathfrak{B})$.

Then for any nuclear operator T *the series*

$$\sum_{j=1}^{\infty} \Delta P_j T G_j \tag{3.2}$$

converges uniformly and the series

$$\sum_{j=1}^{\infty} G_j T \Delta P_j \tag{3.3}$$

converges strongly and absolutely),[1] *while*

$$\left|\sum_{j=1}^{\infty} \Delta P_j T G_j\right| \leq C_1 \sup_j |G_j| \|T\|, \quad \left|\sum_{j=1}^{\infty} G_j T \Delta P_j\right| \leq C_2 \sup_j |G_j| \|T\|, \tag{3.4}$$

where C_1 *and* C_2 *are constants not depending on* T *and* $\{G_j\}_1^\infty$.

Proof. Suppose that T has the form (3.1). Then for any $x \in \mathfrak{B}$

$$\left|\sum_{j=n}^{\infty} \Delta P_j T G_j x\right| \leq \sum_{r=1}^{N} |a_r| \left|\sum_{j=n}^{\infty} \phi_r(G_j x) \Delta P_j \eta_r\right|$$
$$+ \sum_{r=N+1}^{\infty} |a_r| \left|\sum_{j=n}^{\infty} \phi_r(G_j x) \Delta P_j \eta_r\right| \quad (n, N = 1, 2, \cdots).$$

Since, in view of (2.3),

1) We note that the absolute strong convergence of the series (3.3) follows from the fact that every nuclear operator carries an absolutely convergent series of operators into an absolutely convergent series (see for example [11], Russian p. 72).

$$\left|\sum_{j=n}^{\infty}\phi_r(G_jx)\Delta P_j\eta_r\right|\le M\sup_j|G_j\|x|\left|\sum_{j=n}^{\infty}\Delta P_j\eta_r\right|\quad (n,\ r=1,\ 2,\ \cdots)$$

and

$$\left|\sum_{j=n}^{\infty}\Delta P_j\eta_r\right|\le M,$$

we have

$$\left|\sum_{j=n}^{\infty}\Delta P_jTG_jx\right|\le M\sup_j|G_j\|x|\left[\sum_{r=1}^{N}|a_r|\left|\sum_{j=n}^{\infty}\Delta P_j\eta_r\right|+M\sum_{r=N+1}^{\infty}|a_r|\right]$$

$$(N,\ n=1,\ 2,\ \cdots).$$

Hence it immediately follows that the series (3.2) converges, and that the first of relations (3.4) holds with a constant $C_1\le M^2$.

Consider the series (3.3). For it

$$\left|\sum_{j=n}^{\infty}G_jT\Delta P_jx\right|\le\sup_j|G_j|\sum_{r=1}^{\infty}|ar|\sum_{j=n}^{\infty}|\phi_r(\Delta P_jx)| \tag{3.5}$$

$$(x\in\mathfrak{B};\ n=1,\ 2,\ \cdots).$$

As is known (see [5], Russian p. 74), for any vector $x\in\mathfrak{B}$

$$\lim_{n\to\infty}\ \sup_{\phi\in\mathfrak{B}^*,|\phi|=1}\ \sum_{j=n}^{\infty}|\phi(\Delta P_jx)|=0.$$

Accordingly, the series (3.3) converges strongly and absolutely.

The sequence of functionals

$$\mathfrak{P}_n(x)=\sup_{\phi\in\mathfrak{B}^*,|\phi|=1}\ \sum_{j=1}^{n}|\phi(\Delta P_jx)|\quad (x\in\mathfrak{B};\ n=1,\ 2,\ \cdots)$$

satisfies the conditions of the well-known lemma of I. M. Gel′fand (see for example [5], Russian p. 60). Hence, in view of this lemma, for the functional

$$p(x)=\sup_n p_n(x)=\sup_{\phi\in\mathfrak{B}^*,|\phi|=1}\ \sum_{j=1}^{\infty}|\phi(\Delta P_j)|\quad (x\in\mathfrak{B})$$

the relation

$$\sup_{x\in\mathfrak{B}}\frac{P(x)}{|x|}<\infty$$

is satisfied. Hence, using (3.5), we see that the second relation in (3.4) holds.

The lemma is proved.

With Lemma 3 in hand it is easy to obtain the following proposition.

Theorem 4. *Suppose that* T *is a nuclear operator and that* $\|t_{jk}\|_1^\infty$ *is its matrix in some unconditional basis* $\{e_j\}_1^\infty$ *of* $\mathfrak{B}$. *Then for any bounded set* $\{\gamma_{jk}\}$ *of complex numbers the matrix* $\|\gamma_{jk}t_{jk}\|_1^\infty$ *defines in this basis a completely continuous operator* $\Gamma(T)$, *while*

$$|\Gamma(T)| \leq M^2 \sup_{j,k} |\gamma_{jk}| \|T\|,$$

where M *is the constant from* (2.3).

Proof. Consider the discrete chain $\mathfrak{P} = \{P_j\}_0^\infty$, where P_n is a projector projecting the space $\mathfrak{B}$ onto the subspace with the basis $\{e_j\}_1^n$ parallel to the subspace with the basis $\{e_j\}_{n+1}^\infty$. Obviously, the operators

$$G_j = \sum_{k=1}^{\infty} \gamma_{jk} \Delta P_k \quad (j = 1, 2, \dots)$$

are bounded linear operators and

$$|G_j| \leq M \sup_k |\gamma_{jk}|.$$

One verifies immediately that for the operator $\Gamma(T)$ defined in the basis $\{e_j\}_1^\infty$ by the matrix $\|\gamma_{jk}t_{jk}\|_1^\infty$ the equation

$$\Gamma(T) = \sum_{j=1}^{\infty} \Delta P_j T G_j$$

holds, so that, in view of Lemma 3, the validity of Theorem 4 follows.

We shall denote by $\mathfrak{S}_\infty$ the Banach subalgebra of the algebra $\mathfrak{R}(\mathfrak{B})$ consisting of all completely continuous operators, and by $\mathfrak{S}_1$ the Banach algebra of all nuclear operators. On $\mathfrak{S}_1$ we define an operator $\mathfrak{T}$, putting

$$\mathfrak{T}X = \sum_{j=1}^{\infty} P_{j-1} X \Delta P_j \quad (x \in \mathfrak{S}_1).$$

The operator $\mathfrak{T}$ is called a triangular truncation operator, since if $\|x_{jk}\|_1^\infty$ is the matrix of the operator X in the basis $\{e_j\}_1^\infty$ generated by the chain $\mathfrak{P} = \{P_j\}_0^\infty$, then obviously the matrix of the operator $\mathfrak{T}X$ has the form

$\|y_{jk}\|_1^\infty$, where $y_{jk} = 0$ for $j \geq k$ and $y_{jk} = x_{jk}$ for $j < k$.

Theorem 4 implies the following.

Corollary 2. *The operator* $\mathfrak{J}$ *is a bounded linear operator operating from* $\mathfrak{S}_1$ *into* $\mathfrak{S}_\infty$ *and mapping* $\mathfrak{S}_1$ *into the set of Volterra operators of* $\mathfrak{S}_\infty$ *having* $\mathfrak{P}$ *as proper chain.*

Indeed, this proposition follows from Theorem 4 if one puts $\gamma_{jk} = 0$ for $j \geq k$ and $\gamma_{jk} = 1$ for $j < k$.

Many papers have been devoted to the investigation of triangular operators relative to a continuous chain (see the survey [7], and also [8]). The corollary proved above is a discrete analog for Banach space of the original theorem of M. S. Brodskiĭ [9] on the integral of a triangular truncation.

3. From Lemma 3 and Theorem 2 there also immediately results the following theorem.

Theorem 5. *Suppose that* T *is a nuclear operator. Suppose that the operators* $I + P_jTP_j$ *are invertible and* $\sup_j|(I + P_jTP_j)^{-1}| < \infty$. *Then the operator* $A = I + T$ *admits the factorization* (2.2) *relative to the chain* (2.1), *and the operators* $X_\pm$ *and* $D - I$ *are completely continuous.*

4. In this subsection we consider the case when $\mathfrak{B}$ is one of the spaces l_1 and c_0.

Denote by $\widetilde{\mathfrak{S}}_\infty$ the Banach algebra obtained by adjoining the unit operator to the algebra $\mathfrak{S}_\infty$ of all continuous operators.

Suppose that $\mathfrak{P}$ is a discrete chain defined by the standard basis of the space l_1 or c_0. It is easy to see that in this case each operator $A = I + T$ $(T \in \mathfrak{S}_\infty)$ is representable in the form $A = A_- + A_+$, where

$$A_+ = \sum_{j=1}^{\infty} P_{j-1} T \Delta P_j \tag{3.6}$$

is an operator of $\mathfrak{S}_\infty$ having the proper chain $\mathfrak{P}$, and the operator

$$A_- = I + \sum_{j=1}^{\infty} \Delta P_j T P_j \tag{3.7}$$

differs from the unit operator by a completely continuous operator and has as proper chain the chain $\mathfrak{Q}$ complementary to $\mathfrak{P}$,

Moreover, $|A_\pm| \leq |A|$.

Denote by $\widetilde{\mathfrak{S}}_\infty^+$ the set of operators of the form (3.6) and by $\widetilde{\mathfrak{S}}_\infty^-$ the set of operators of the form (3.7). It is easy to see that $\widetilde{\mathfrak{S}}_\infty^\pm$ form closed subalgebras of the algebra $\widetilde{\mathfrak{S}}_\infty$ and that their direct sum is equal to $\widetilde{\mathfrak{S}}_\infty$. This fact is fundamental in the following theorem.

Theorem 6. *In the space* $\mathfrak{B} = l_1, c_0$, *any invertible operator of the form* $A = I + T$, $T \in \mathfrak{S}_\infty$, *satisfying condition* a), *admits factorization relative to the chain* $\mathfrak{P}$ *defined by the standard basis, while the operators* $X_\pm - I$ *and* $D - I$ *are in* $\mathfrak{S}_\infty$.

Proof. Represent the operator $A = I + T$ in the form

$$I + T = (I + K)(I + M), \tag{3.8}$$

where K is a finite-dimensional operator and M is a continuous operator, sufficiently small in norm. In view of Lemma 5.1 of [1] the operator $I + M$ may be represented in the form

$$I + M = (I + M_-)(I + M_+), \tag{3.9}$$

where $I + M_\pm$ are invertible operators and $M_\pm \in \widetilde{\mathfrak{S}}_\infty^\pm$. Then

$$A = Y_-(I + Y_-^{-1}KY_-)Y_+ \tag{3.10}$$

where $Y_\pm = I + M_\pm$.

It follows from (3.10) that the operator $I + Y_-^{-1}KY_-$ satisfies condition a), and therefore, by Theorem 5, it admits factorization relative to the chain $\mathfrak{P}$:

$$I + Y_-^{-1}KY_- = Z_-\widetilde{Z}_+ \ (\widetilde{Z}_+ = D'Z_+). \tag{3.11}$$

As a result we find for the operator A the equation $A = Y_-Z_-\widetilde{Z}_+Y_+$, or

$$A = X_-DX_+, \tag{3.12}$$

where $X_\pm \in \mathfrak{R}_\pm(\mathfrak{P})$, $D \in \mathfrak{R}_0(\mathfrak{P})$ and $\Delta P_j X_\pm \Delta P_j = \Delta P_j$ $(j = 1, 2, \cdots)$, while the operators $X_\pm - I$ and $D - I$ are completely continuous.

The theorem is proved.

As in §2, all the propositions of this section remain in force also for the chain $\mathfrak{Q}$ complementary to $\mathfrak{P}$. We note that Theorem 6 for this case was proved in the paper [10].

5. We consider now the case of Hilbert space $\mathfrak{H}$. Suppose that $\mathfrak{P} = \{P_j\}_0^\infty$ is a discrete chain of orthogonal projectors of $\mathfrak{R}(\mathfrak{H})$. Then we have the following.

Theorem 7. *Suppose that the invertible operator* A *satisfies condition*

a) *and has the form* $A = H + T$, *where* H *is a positive invertible operator,*[1] *and* $T \in \mathfrak{S}_\omega$.[2] *Then the operator* A *admits the factorization* (2.2) *relative to the chain* $\mathfrak{P} = \{P_j\}_0^\infty$.

Proof. Since H is a positive invertible operator, there exists a $\delta > 0$ such that

$$(Hx, x) \geq \delta(x, x) \quad (x \in \mathfrak{H}), \tag{3.13}$$

so that

$$(P_j H P_j x, x) \geq \delta(x, x) \quad (x \in P_j \mathfrak{H}). \tag{3.14}$$

It follows from (3.14) that the operators $P_j H P_j$ $(j = 1, 2, \cdots)$ are invertible in the corresponding subspaces $P_j \mathfrak{H}$, and $\sup_j |(P_j H P_j)^{-1}| < \infty$.

Thus the operator H satisfies condition a), and, from Lemma 1, the series (2.7) and (2.10) converge for it.

In view of Theorem 1 the operator $H_n = P_n H P_n$ admits the factorization

$$H_n = Y_{n-} D_n Y_{n+} \tag{3.15}$$

relative to the chain $\mathfrak{P}_n = \{P_j\}_0^n$ in the subspace $P_n \mathfrak{H}$. Since the equation $H_n = Y_{n+}^* D_n^* Y_{n-}^*$ also yields a factorization of the operator H_n relative to the same chain, then $Y_{n-} = Y_{n+}^*$, $D_n^* = D_n > 0$ and equation (3.15) may be written in the form

$$H_n = \widetilde{Y}_{n+}^* \widetilde{Y}_{n+} \quad (\widetilde{Y}_{n+} = D_n^{1/2} Y_{n+}), \tag{3.16}$$

from which it follows that $|Y_{n-}| = |Y_{n+}| \leq |D_n^{-1/2}| \, \|H_n|^{1/2}$ and

$$\sup_n |Y_{n+}| < \infty. \tag{3.17}$$

Taking account of the equation $P_n Y_{n\pm} P_n = Y_{n\pm}$ $(n = 1, 2, \cdots)$, we find that the sequence $\{y_{n\pm}\}_1^\infty$ converges on the set $\bigcup_j P_j \mathfrak{H}$, dense in $\mathfrak{H}$. Therefore we conclude from (3.17) that this sequence converges everywhere. Thus, in view of Theorem 2, the operator H admits the factorization

$$H = Y_- \widetilde{Y}_+ \quad (\widetilde{Y}_+ = D' Y_+) \tag{3.18}$$

---

1) In place of the positivity and invertibility of the operator H it suffices to require that it be nonnegative.

2) For the definition of the class $\mathfrak{S}_\omega$ see [1].

relative to the chain $\mathfrak{P} = \{P_j\}_0^\infty$.

The operator A may be represented in the form

$$A = Y_-(I + Y_-^{-1} T \widetilde{Y}_+^{-1}) \widetilde{Y}_+. \tag{3.19}$$

It follows from this equation that the operator $I + Y_-^{-1} T \widetilde{Y}_+^{-1} = I + T_1$ is invertible and satisfies condition a). Since moreover the operator $T_1 \in \mathfrak{S}_\omega$, the operator $I + T_1$ (see [1], Theorems 6.2 and 7.2) admits the factorization

$$I + T_1 = Z_- \widetilde{Z}_+ \quad (\widetilde{Z}_+ = D'' Z_+). \tag{3.20}$$

It follows from (3.19) and (3.20) that the operator A satisfies the equation

$$A = Y_- Z_- \widetilde{Z}_+ \widetilde{Y}_+, \tag{3.21}$$

equivalent to its factorization (2.2).

The theorem is proved.

The following is an immediate consequence.

Corollary 3. *If an invertible selfadjoint operator A satisfies condition* a) *and all of its negative spectrum consists of a finite number of eigenvalues of finite multiplicity, then it admits the factorization* (2.2) *relative to the chain* $\mathfrak{P} = \{P_j\}_0^\infty$.

An easy consequence of this is the following.

Corollary 4. *A selfadjoint invertible operator $A = I + T$ $(T \in \mathfrak{S}_\infty)$ satisfying condition* a) *admits the factorization* (2.2) *relative to the chain* $\mathfrak{P} = \{P_j\}_0^\infty$.

We note further that all of the propositions of §§2 and 3 generalize in a natural way for chains of the form $\mathfrak{P} = \{P_j\}_{-\infty}^\infty$, for which

$$\dim \Delta P_j \mathfrak{B} = 1 \tag{3.22}$$

and the series

$$\sum_{j=-\infty}^{\infty} \Delta P_j x \tag{3.23}$$

converge unconditionally for any vector $x \in \mathfrak{B}$.

Condition (3.22) may be dropped if one imposes certain additional restrictions on the chain $\mathfrak{P}$.

BIBLIOGRAPHY

[1] I. C. Gohberg and M. G. Kreĭn, *Factorization of operators in Hilbert space,* Acta Sci. Math. Szeged 25 (1964), 90–123; English transl., Amer. Math. Soc. Transl. (2) 51 (1966), 155–188. MR 29 #6313.

[2] M. Shinbrot, *On singular integral operators,* J. Math. Mech. 13 (1964), 395–406. MR 28 #5304.

[3] M. M. Grinbljum, *On the representation of a space of type B in the form of a direct sum of subspaces,* Dokl. Akad. Nauk SSSR 70 (1950), 749–752. (Russian) MR 11, 525.

[4] I. C. Gohberg and M. G. Kreĭn, *Introduction to the theory of linear nonselfadjoint operators in Hilbert space,* "Nauka", Moscow, 1965; English transl., Transl. Math. Monographs, vol. 18, Amer. Math. Soc., Providence, R. I., 1969. MR 36 #3137.

[5] E. Hille and R. Phillips, *Functional analysis and semigroups,* rev. ed., Amer. Math. Soc. Colloq. Publ., vol. 31, Amer. Math. Soc., Providence, R. I., 1957; Russian transl., IL, Moscow, 1962. MR 19, 664.

[6] N. I. Ahiezer and I. M. Glazman, *The theory of linear operators in Hilbert space,* GITTL, Moscow, 1950; English transl., Ungar, New York, 1961. MR 13, 358.

[7] M. S. Brodskiĭ, I. C. Gohberg, M. G. Kreĭn and V. I. Macaev, *Some investigations in the theory of non-selfadjoint operators,* Proc. Fourth All-Union Math. Congress, vol. II: Sectional Lectures, "Nauka", Leningrad, 1964, pp. 261–271; English transl., Amer. Math. Soc. Transl. (2) 65 (1967), 237–251. MR 36 #3153.

[8] I. C. Gohberg and M. G. Kreĭn, *Theory and applications of Volterra operators in Hilbert space,* "Nauka", Moscow, 1967; English transl., Transl. Math. Monographs, vol. 24, Amer. Math. Soc., Providence, R.I., 1970. MR 36 #2007.

[9] M. S. Brodskiĭ, *On the triangular representation of completely continuous operators with one-point spectra,* Uspehi Mat. Nauk 16 (1961), no. 1 (97), 135–141; English transl., Amer. Math. Soc. Transl. (2) 47 (1965), 59–65. MR 24 #A426.

[10] G. Strang, *Wiener-Hopf difference equations,* J. Math. Mech. 13 (1964), 85–96. MR 28 #3548.

[11] B. S. Mitjagin, *Approximate dimension and bases in nuclear spaces*, Uspehi Mat. Nauk **16** (1961), no. 4 (100), 63–132 = Russian Math. Surveys 16 (1961), no. 4, 59–127. MR 27 #2837.

Translated by:
J. M. Danskin

Amer. Math. Soc. Transl.
(2) Vol. 90, 1970

ON THE FACTORIZATION OF OPERATORS IN A BANACH SPACE*

M. A. BARKAR′ AND I. C. GOHBERG

The abstract problem of factorizing operators with respect to a chain of orthogonal projectors was formulated and solved in a paper of M. G. Kreĭn and I. C. Gohberg [1] for a wide class of operators acting in Hilbert space.

In [2] some of these results are extended to operators acting in a Banach space for the case of a discrete chain of projectors.

In the present paper we obtain extensions of some theorems of [1] for the general case of a continuous chain in a Banach space.

The paper consists of four sections. In the first section we prove general theorems on the factorization of operators with respect to a continuous chain. As in the case of Hilbert space [1], these theorems reduce the problem of factorizing an operator to a question of the convergence of certain integrals over a chain. In the second section we consider classes of chains that are important for the sequel (chains of bounded vector and of bounded operator variation). In the third section we prove the convergence of certain integrals over chains, and in the fourth section we establish that, for certain classes of chains (for example, for chains of bounded operator variation), an operator differing from the identity operator by a nuclear operator admits a factorization.

The authors wish to express their gratitude to A. S. Markus for a useful discussion of the results of this paper.

§1. General theorems on the factorization of an operator with respect to a chain

1. Let $\mathfrak{R}$ denote the ring of all bounded linear operators acting in a given Banach space $\mathfrak{B}$.

We introduce a partial ordering in the set of projectors in $\mathfrak{R}$ by putting $P_1 < P_2$ if $P_1\mathfrak{B} \subset P_2\mathfrak{B}$ and $(I - P_1)\mathfrak{B} \supset (I - P_2)\mathfrak{B}$.

A set $\mathfrak{P} = \{P\}$ of projectors from $\mathfrak{R}$ is called a chain if it is ordered and contains the zero and identity operators. The chain $\mathfrak{Q} = \{Q\}$ $(Q = I - P,\ P \in \mathfrak{P})$ is called the complement of $\mathfrak{P}$.

---

* Translation of Mat. Issled. 1 (1966), no. 2, 98–129.

A chain $\mathfrak{P} = \{P\}$ is called an eigenchain of an operator A if, for any projector $P \in \mathfrak{P}$, the subspace $P\mathfrak{B}$ is invariant with respect to A, i.e. $AP = PAP$.

Let $\mathfrak{R}_+(\mathfrak{P})$ denote the ring of operators from $\mathfrak{R}$ for which $\mathfrak{P}$ is an eigenchain, let $\mathfrak{R}_-(\mathfrak{P})$ denote the ring $\mathfrak{R}_+(\mathfrak{Q})$ and let $\mathfrak{R}_0(\mathfrak{P})$ denote the intersection $\mathfrak{R}_+(\mathfrak{P}) \cap \mathfrak{R}_-(\mathfrak{P})$. Clearly, $\mathfrak{R}_0(\mathfrak{P})$ consists of all operators $A \in \mathfrak{R}$ commuting with any projector $P \in \mathfrak{P}$.

A chain $\mathfrak{Z} = \{0 = P_0 < P_1 < \cdots < P_n = I\}$ consisting of a finite number of projectors from $\mathfrak{P}$ is called a partitioning of $\mathfrak{P}$.

We will say that an invertible operator A admits a factorization relative to a chain $\mathfrak{P} = \{P\}$ if it is representable in the form of the product

$$A = X_- D X_+ \tag{1.1}$$

with the factors satisfying the following conditions: the operators $X_\pm$ are invertible,

$$X_\pm \in \mathfrak{R}_\pm(\mathfrak{P}), \; X_\pm^{-1} \in \mathfrak{R}_\pm(\mathfrak{P}), \; D \in \mathfrak{R}_0(\mathfrak{P}) \tag{1.2}$$

and

$$\int_{\mathfrak{P}} dPX_\pm dP = I. \tag{1.3}$$

Let us explain the latter condition. For an arbitrary operator $A \in \mathfrak{R}$ we will say that the integral $\int_{\mathfrak{P}} dPAdP$ converges and is equal to an operator B if, for any $\epsilon > 0$, a partitioning $\mathfrak{Z}_\epsilon$ of $\mathfrak{P}$ can be found such that for any partitioning $\mathfrak{Z} = \{P_j\}_0^n \supset \mathfrak{Z}_\epsilon$

$$\left| \sum_{j=1}^{n} (P_j - P_{j-1}) A (P_j - P_{j-1}) - B \right| < \epsilon.$$

One can easily see that if an invertible operator A admits the representation (1.1) then the operator D has an inverse $D^{-1} \in \mathfrak{R}_0(\mathfrak{P})$ and

$$\int dPX_\pm^{-1} dP = I. \tag{1.4}$$

1°. *If an invertible operator A admits a factorization relative to a chain $\mathfrak{P}$ then the factors $X_\pm$ and D are uniquely defined.*

For suppose

$$A = X_- D X_+ \text{ and } A = Y_- D_1 Y_+, \tag{1.5}$$

with

$$X_{\pm},\ Y_{\pm} \in \mathfrak{R}_{\pm}(\mathfrak{P});\ X_{\pm}^{-1}Y_{\pm}^{-1} \in \mathfrak{R}_{\pm}(\mathfrak{P});\ D,\ D_1 \in \mathfrak{R}_0(\mathfrak{P}) \tag{1.6}$$

and

$$\int_{\mathfrak{P}} dPX_{\pm}dP = \int_{\mathfrak{P}} dPY_{\pm}dP = I. \tag{1.7}$$

Then

$$Y_+X_+^{-1} = D_1^{-1}Y_-^{-1}X_-D.$$

From this equality we get by virtue of (1.6) that for an arbitrary partitioning $\mathfrak{Z} = \{P_j\}_0^n$ of $\mathfrak{P}$

$$Y_+X_+^{-1} = \sum_{j=1}^{n} \Delta P_j Y_+X_+^{-1}\Delta P_j \quad (\Delta P_j = P_j - P_{j-1})$$

or

$$Y_+X_+^{-1} = \sum_{j=1}^{n} \Delta P_j Y_+ \Delta P_j \sum_{j=1}^{n} \Delta P_j X_+^{-1} \Delta P_j$$

and hence, according to (1.4) and (1.7),

$$Y_+X_+^{-1} = I,\ Y_+ = X_+.$$

Similarly, it is seen that $Y_- = X_-$ and hence $D_1 = D$.

The assertion is proved.

Theorem 1. *If an invertible operator A admits a factorization relative to a chain $\mathfrak{P} \in \{P\}$, then*

a) *for any projector $P \in \mathfrak{P}$ the operator PAP is invertible in the subspace $P\mathfrak{B}$;*

b) *the factors $X_{\pm}$ and D of this factorization are expressed by the formulas*

$$X_- = \overset{m}{\overleftarrow{\int_{\mathfrak{P}}}} (I + dPAP(PAP)^{-1}), \tag{1.8}$$

$$X_+ = \overset{m}{\overrightarrow{\int_{\mathfrak{P}}}} (I + (PAP)^{-1}PAdP), \tag{1.9}$$

$$D = \int_{\mathfrak{P}}^{m} dP(A - AP(PAP)^{-1}PA)dP, \tag{1.10}$$

and their inverses, by the formulas

$$X_{-}^{-1} = I - \int_{\mathfrak{P}}^{m} dPAP(PAP)^{-1}, \tag{1.11}$$

$$X_{+}^{-1} = I - \int_{\mathfrak{P}}^{m} (PAP)^{-1}PAdP, \tag{1.12}$$

$$D^{-1} = \int_{\mathfrak{P}}^{M} dP(PAP)^{-1}dP. \tag{1.13}$$

Conversely, if condition a) *is satisfied and the three integrals* (1.8)–(1.10), *or* (1.11)–(1.13), *converge for an invertible operator A then A admits the factorization* (1.1) *relative to $\mathfrak{P}$, and consequently the factors of this factorization admit the representations* (1.8)–(1.13).

2. Let us define the integrals over a chain which appear in this theorem. Let $\Phi(\mathfrak{Z})$ be an arbitrary, in general multivalued, operator function defined on all of the partitionings $\mathfrak{Z}$ of a chain $\mathfrak{P}$ with values in $\mathfrak{R}$.

An operator A is said to be the linit of $\Phi(\mathfrak{Z})$ with respect to $\mathfrak{P}$ (in the sense of S. O. Šatunovskiĭ):

$$A = \lim \Phi(\mathfrak{Z}),$$

if for any $\epsilon > 0$ there exists a partitioning $\mathfrak{Z}_\epsilon$ of $\mathfrak{P}$ such that

$$|\Phi(\mathfrak{Z}) - A| < \epsilon$$

for all extended partitionings $\mathfrak{Z}$ $(\supset \mathfrak{Z}_\epsilon)$,

Consider arbitrary operator functions $F(P)$ and $G(P)$ defined on a chain $\mathfrak{P} = \{P\}$ with values in $\mathfrak{R}$. An operator function $S(\mathfrak{Z})$ can be formed from them by putting

$$S(\mathfrak{Z}) = \sum_{j=1}^{n} F(P_{j-1})\Delta P_j G(P_{j-1}) \tag{1.14}$$

for each partitioning $\mathfrak{Z} = \{P_j\}_0^n$ of $\mathfrak{P}$.

If $S(\mathfrak{Z})$ has an operator $A \in \mathfrak{R}$ as its limit, we will say that the integral

$$\int_{\mathfrak{P}}^{m} F(P)dPG(P) \tag{1.15}$$

converges uniformly (or, simply, converges) and is equal to A.

The multiplicative integrals

$$\overset{m}{\overrightarrow{\int_{\mathfrak{P}}}}(I + F(P)dPG(P)), \quad \overset{m}{\overleftarrow{\int_{\mathfrak{P}}}}(I + F(P)dPG(P)) \tag{1.16}$$

and the integral

$$\overset{m}{\int_{\mathfrak{P}}} dPF(P)dP \tag{1.17}$$

are defined as the limits of the corresponding functions

$$\overrightarrow{\prod}(\mathfrak{Z}) = \overrightarrow{\prod_{j=1}^{n}}(I + F(P_{j-1})\Delta P_j G(P_{j-1})), \tag{1.18}$$

$$\overleftarrow{\prod}(\mathfrak{Z}) = \overleftarrow{\prod_{j=1}^{n}}(I + F(P_{j-1})\Delta P_j G(P_{j-1})), \tag{1.19}$$

$$\hat{S}(\mathfrak{Z}) = \sum_{j=1}^{n} \Delta P_j F(P_{j-1})\Delta P_j. \tag{1.20}$$

If in the definition of the integrals (1.15)–(1.17) we replace the equalities (1.14) and (1.18)–(1.20) for the functions $S(\mathfrak{Z})$, $\overrightarrow{\Pi}(\mathfrak{Z})$, $\overleftarrow{\Pi}(\mathfrak{Z})$ and $\hat{S}(\mathfrak{Z})$ by the equalities

$$S(\mathfrak{Z}) = \sum_{j=1}^{n} F(P_j)\Delta P_j G(P_j), \tag{1.21}$$

$$\overrightarrow{\prod}(\mathfrak{Z}) = \overrightarrow{\prod_{j=1}^{n}}(I + F(P_j)\Delta P_j G(P_j)), \tag{1.22}$$

$$\overleftarrow{\prod}(\mathfrak{Z}) = \overleftarrow{\prod_{j=1}}(I + F(P_j)\Delta P_j G(P_j)), \tag{1.23}$$

$$\hat{S}(\mathfrak{Z}) = \sum_{j=1}^{n} \Delta P_j F(P_j)\Delta P_j, \tag{1.24}$$

we arrive at a definition of the integrals

$$\int_{\mathfrak{P}}^{M} F(P)\,dPG(P), \quad \overrightarrow{\int_{\mathfrak{P}}^{M}} (I + F(P)\,dPG(P)),$$

$$\overleftarrow{\int_{\mathfrak{P}}^{M}} (I + F(P)\,dPG(P), \quad \int_{\mathfrak{P}}^{M} dPF(P)\,dP. \tag{1.25}$$

3. **Proof of the theorem.** Suppose the operator A admits the factorization (1.1) relative to the chain $\mathfrak{P}$.

Then the operators $PX_{\pm}P$ are invertible in the subspace $P\mathfrak{B}$, and $(PX_{\pm}P)^{-1} = PX_{\pm}^{-1}P$. The equality

$$PAP = (PX_{-}P)D(PX_{+}P)$$

implies that the operator PAP is also invertible in this subspace.

It follows from (1.1) by virtue of (1.2) that

$$(P_{j-1}AP_{j-1})^{-1}P_{j-1}A\Delta P_j = X_{+}^{-1}P_{j-1}X_{+}\Delta P_j$$

and, since $P_{j-1} = P_j - \Delta P_j$, that

$$(P_{j-1}AP_{j-1})^{-1}P_{j-1}A\Delta P_j = \Delta P_j - X_{+}^{-1}\Delta P_j X_{+}\Delta P_j.$$

Consequently, if $\mathfrak{Z} = \{P_j\}_0^n$ is a partitioning of $\mathfrak{P}$, then

$$\sum_{j=1}^{n} (P_{j-1}AP_{j-1})^{-1}P_{j-1}A\Delta P_j = I - X_{+}^{-1}\sum_{j=1}^{n}\Delta P_j X_{+}\Delta P_j \tag{1.26}$$

from which, according to (1.3), we obtain equality (1.12).

Noting that

$$\left[\overrightarrow{\prod_{j=1}^{n}}(I + (P_{j-1}AP_{j-1})^{-1}P_{j-1}A\Delta P_j\right]^{-1} = I - \sum_{j=1}^{n}(P_{j-1}AP_{j-1})^{-1}P_{j-1}A\Delta P_j,$$

we find from (1.26) that

$$\overrightarrow{\prod_{j=1}^{n}}(I + (P_{j-1}AP_{j-1})^{-1}P_{j-1}A\Delta P_j) = \left[\sum_{j=1}^{n}\Delta P_j X_{+}^{-1}\Delta P_j\right]X_{+}, \tag{1.27}$$

and this equality and (1.4) imply equality (1.9).

The remainder of formulas (1.8)–(1.13) are proved analogously.

Suppose now condition a) is satisfied and the integrals (1.8)–(1.10) converge to the operators X_-, X_+ and D respectively.

If $\mathfrak{Z} = \{P_j\}_0^n$ is an arbitrary parititioning of $\mathfrak{P}$, then, by virtue of Theorem 1 in [2], the operator A admits a factorization with respect to the finite chain $\mathfrak{Z}$:

$$A = X_-(\mathfrak{Z})D(\mathfrak{Z})X_+(\mathfrak{Z}), \tag{1.28}$$

where

$$X_-(\mathfrak{Z}) = \overleftarrow{\prod_{j=1}^{n}} (I + \Delta P_j A P_{j-1}(P_{j-1}AP_{j-1})^{-1}),$$

$$X_+(\mathfrak{Z}) = \overrightarrow{\prod_{j=1}^{n}} (I + (P_{j-1}AP_{j-1})^{-1}P_{j-1}A\Delta P_j),$$

$$D(\mathfrak{Z}) = \sum_{j=1}^{n} \Delta P_j(A - AP_{j-1}(P_{j-1}AP_{j-1})^{-1}P_{j-1}A)\Delta P_j,$$

and

$$X_-^{-1}(\mathfrak{Z}) = I - \sum_{j=1}^{n} \Delta P_j A P_{j-1}(P_{j-1}AP_{j-1})^{-1},$$

$$X_+^{-1}(\mathfrak{Z}) = I - \sum_{j=1}^{n} (P_{j-1}AP_{j-1})^{-1}P_{j-1}A\Delta P_j,$$

$$D^{-1}(\mathfrak{Z}) = \sum_{j=1}^{n} \Delta P_j(P_jAP_j)^{-1}\Delta P_j,$$

with, as is easily seen,

$$X_\pm(\mathfrak{Z}) \in \mathfrak{R}_\pm(\mathfrak{P});\ X_\pm^{-1}(\mathfrak{Z}) \in \mathfrak{R}_\pm(\mathfrak{P});\ D(\mathfrak{Z}),\ D^{-1}(\mathfrak{Z}) \in \mathfrak{R}_0(\mathfrak{Z}). \tag{1.29}$$

Since

$$\lim X_\pm(\mathfrak{Z}) = X_\pm \text{ and } \lim D(\mathfrak{Z}) = D,$$

it follows from equality (1.28) that

$$A = X_-DX_+, \tag{1.30}$$

in which, by virtue of (1.29),

$$X_\pm \in \mathfrak{R}_\pm(\mathfrak{P}),\ D \in \mathfrak{R}_0(\mathfrak{P}). \tag{1.31}$$

Equality (1.30) shows that the operators X_+ and X_- are invertible from the left and right respectively. But then it is easily seen that these operators are invertible and

$$\lim X_\pm^{-1}(\mathfrak{Z}) = X_\pm^{-1}.$$

From this result and (1.29) it follows that

$$X_\pm^{-1} \in \mathfrak{R}_\pm(\mathfrak{P}). \tag{1.32}$$

Taking the convergence of the integral (1.9) into account, we get from equality (1.27) that

$$\int_{\mathfrak{P}} dP X_+^{-1} dP = I. \tag{1.33}$$

It is established in exactly the same way that

$$\int_{\mathfrak{P}} dP X_-^{-1} dP = I. \tag{1.34}$$

Relations (1.31)–(1.34) show that equality (1.30) gives the factorization of A relative to $\mathfrak{P}$.

The sufficiency of the conditions of the theorem in the case of the convergence of the integrals (1.11)–(1.13) is proved analogously.

The theorem is proved.

4. We note that we have simultaneously established a connection between additive and multiplicative integrals over a chain of projectors:

$$\left[\overset{m}{\overleftarrow{\int_{\mathfrak{P}}}} (I + dPAP(PAP)^{-1}\right]^{-1} = I - \int_{\mathfrak{P}} dPAP(PAP)^{-1}, \tag{1.35}$$

$$\left[\overset{m}{\overrightarrow{\int_{\mathfrak{P}}}} (I + (PAP)^{-1}PAdP)\right]^{-1} = I - \overset{m}{\int_{\mathfrak{P}}} (PAP)^{-1}PAdP, \tag{1.36}$$

as well as the equality

$$\left[\overset{m}{\int_{\mathfrak{P}}} dP(A - AP(PAP)^{-1}PA)dP\right]^{-1} = \overset{M}{\int_{\mathfrak{P}}} dP(PAP)^{-1}dP. \tag{1.37}$$

In addition to this theorem we show that if an invertible operator A admits a factorization relative to a chain $\mathfrak{P}$, then the integrals

$$\int_{\mathfrak{P}}^{M} (PAP)^{-1}dP \text{ and } \int_{\mathfrak{P}}^{M} dP(PAP)^{-1}$$

converge and the following equalities hold:

$$\int_{\mathfrak{P}}^{M} (PAP)^{-1}dP = X_{+}^{-1}D^{-1} \tag{1.38}$$

and

$$\int_{\mathfrak{P}}^{M} dP(PAP)^{-1} = D^{-1}X_{-}^{-1}. \tag{1.39}$$

For suppose $A = X_{-}\widetilde{X}_{+}$ $(\widetilde{X}_{+} = DX_{+})$, with conditions (1.2) and equality (1.3) being satisfied. Then

$$(P_jAP_j)^{-1}\Delta P_j = \widetilde{X}_{+}^{-1}P_jX_{-}^{-1}\Delta P_j.$$

Replacing P_j by $\Delta P_j + P_{j-1}$ in the right side of this equality and taking into account the fact that $P_{j-1}X_{-}^{-1}\Delta P_j = 0$, we obtain

$$(P_jAP_j)^{-1}\Delta P_j = \widetilde{X}_{+}^{-1}\Delta P_jX_{-}^{-1}\Delta P_j.$$

If $\mathfrak{Z} = \{P_j\}_0^n$ is a partitioning of $\mathfrak{P}$, then

$$\sum_{j=1}^{n}(P_jAP_j)^{-1}\Delta P_j = \widetilde{X}_{+}^{-1}\sum_{j=1}^{n}\Delta P_jX_{-}^{-1}\Delta P_j,$$

which implies by virtue of (1.4) that the integral

$$\int_{\mathfrak{P}}^{M} (PAP)^{-1}dP$$

converges and equality (1.38) holds.

The convergence of the integral

$$\int_{\mathfrak{P}}^{M} dP(PAP)^{-1}$$

and the validity of equality (1.39) are proved analogously.

5. We present some additions to Theorem 1 for the case when the operator A has the form $A = I + T$, where $T \in \mathfrak{S}_\infty$. By $\mathfrak{S}_\infty$ we denote the set of completely continuous operators of the ring $\mathfrak{R}$. In this case formulas (1.8)–(1.10) are equivalent to

$$X_- = \overset{m}{\overleftarrow{\int_{\mathfrak{P}}}} (I + dPTP(I + PTP)^{-1}), \tag{1.40}$$

$$X_+ = \overset{m}{\overrightarrow{\int_{\mathfrak{P}}}} (I + (I + PTP)^{-1} PTdP), \tag{1.41}$$

$$D = I + \int_{\mathfrak{P}}^{m} dP(T - TP(I + PTP)^{-1} PT)dP, \tag{1.42}$$

while formulas (1.11)–(1.13) are equivalent to

$$X_-^{-1} = I - \int_{\mathfrak{P}}^{m} dPTP(I + PTP)^{-1}, \tag{1.43}$$

$$X_+^{-1} = I - \int_{\mathfrak{P}}^{m} (I + PTP)^{-1} PTdP, \tag{1.44}$$

$$D^{-1} = I - \int_{\mathfrak{P}}^{M} dPTP(I + PTP)^{-1} dP. \tag{1.45}$$

Equalities (1.38) and (1.39) are written as follows:

$$X_+^{-1} D^{-1} = I - \int_{\mathfrak{P}}^{M} (I + PTP)^{-1} PTdP, \tag{1.46}$$

$$D^{-1} X_-^{-1} = I - \int_{\mathfrak{P}}^{M} dPTP(I + PTP)^{-1}. \tag{1.47}$$

These formulas imply that if an operator $A = I + T$ $(T \in \mathfrak{S}_\infty)$ admits the factorization (1.1) relative to a chain $\mathfrak{P}$, then the factors $X_\pm$ and D of this factorization also differ from the identity operator by a completely continuous operator.

Theorem 2. *Suppose an operator* $A = I + T$ $(T \in \mathfrak{S}_\infty)$ *satisfies condition* a) *and one of the integrals* (1.42) *or* (1.45) *converges.*

If at least one of the integrals (1.40), (1.41), (1.43), (1.44), (1.46) *or* (1.47) *converges then the operator* A *admits the factorization* (1.1) *relative to the chain* $\mathfrak{P}$, *and consequently formulas* (1.40)–(1.47) *hold for the factors of this factorization.*

Proof. Let $\mathfrak{Z} = \{P_j\}_0^n$ be an arbitrary partitioning of the chain $\mathfrak{P}$. By virtue of Theorem 1 in [2] the operator A admits a factorization relative to the chain $\mathfrak{Z} = \{P_j\}_0^n$:

$$A = X_-(\mathfrak{Z})D(\mathfrak{Z})X_+(\mathfrak{Z}), \tag{1.48}$$

where

$$X_-(\mathfrak{Z}) = \overset{\curvearrowleft}{\prod_{j=1}^{n}} (I + \Delta P_j T P_{j-1}(I + P_{j-1} T P_{j-1})^{-1}),$$

$$X_+(\mathfrak{Z}) = \overset{\curvearrowright}{\prod_{j=1}^{n}} (I + (I + P_{j-1} T P_{j-1})^{-1} P_{j-1} T \Delta P_j),$$

$$D(\mathfrak{Z}) = I + \sum_{j=1}^{n} \Delta P_j (T - T P_{j-1}(I + P_{j-1} T P_{j-1})^{-1} P_{j-1} T) \Delta P_j$$

and

$$X_-^{-1}(\mathfrak{Z}) = I - \sum_{j=1}^{n} \Delta P_j T P_{j-1}(I + P_{j-1} T P_{j-1})^{-1},$$

$$X_+^{-1}(\mathfrak{Z}) = I - \sum_{j=1}^{n} (I + P_{j-1} T P_{j-1})^{-1} P_{j-1} T \Delta P_j,$$

$$D^{-1}(\mathfrak{Z}) = I - \sum_{j=1}^{n} \Delta P_j T P_j (I + P_j T P_j)^{-1} \Delta P_j.$$

If, for example, the integrals (1.40) and (1.42) converge to the operators X_- and D respectively, then, rewriting equality (1.48) in the form

$$X_+^{-1}(\mathfrak{Z}) = A^{-1} X_-(\mathfrak{Z}) D(\mathfrak{Z}),$$

we arrive at the conclusion that the integral (1.44) converges.

But, inasmuch as its nth power is equal to zero, the operator

$$\sum_{j=1}^{n} (I + P_{j-1} T P_{j-1})^{-1} P_{j-1} T \Delta P_j,$$

is a Volterra operator.[1)]

Since the uniform limit of a sequence of Volterra operators is also a Volterra operator it follows that $\lim X_+^{-1}(\mathfrak{Z})$ has an inverse; but then, as is easily seen, $\lim X_+(\mathfrak{Z})$ exists and by virtue of Theorem 1 the operator $A = I + T$ $(T \in \mathfrak{S}_\infty)$ admits the factorization (1.1) relative to $\mathfrak{P}$.

The sufficiency of the conditions of the theorem in the remaining cases is proved analogously.

The theorem is proved.

Corollary. *If an operator $A = I + T$ $(T \in \mathfrak{S}_\infty)$ satisfies condition* a) *and either the integrals* (1.40) *and* (1.41) *or the integrals* (1.43) *and* (1.44) *converge, then A admits a factorization relative to the chain $\mathfrak{P}$.*

For suppose an operator $A = I + T$ $(T \in \mathfrak{S}_\infty)$ satisfies condition a) and the integrals (1.40) and (1.41), for example, converge. Then, as in the proof of the theorem, we see that the integrals (1.43) and (1.44) also converge, and from equality (1.48) it follows that the integral (1.42) converges.

§ 2. Chains of bounded variation and their properties

1. A chain $\mathfrak{P}$ is said to be closed if any monotone sequence of projectors from $\mathfrak{P}$ strongly converges to a projector from $\mathfrak{P}$. Clearly, every closed chain of projectors is bounded.

We will assume below without saying so explicitly that all chains are closed.

A pair of projectors (P^-, P^+) $(P^- < P^+, P^\pm \in \mathfrak{P})$ is called a gap of $\mathfrak{P}$ if there is no projector in $\mathfrak{P}$ situated between P^- and P^+. By the dimension of a gap we will mean the dimension of the subspace $(P^+ - P^-)\mathfrak{B}$. A chain without gaps is said to be continuous.

A chain $\mathfrak{P} = \{P\}$ is called a chain of bounded vector variation if for any vector $x \in \mathfrak{B}$ and any functional $\phi \in \mathfrak{B}^*$

$$\sup_{\mathfrak{Z}} \sum_{j=1}^{n} |\phi(\Delta P_j x)| < \infty, \tag{2.1}$$

where the supremum is taken over all of the partitionings $\mathfrak{Z}$ of $\mathfrak{P}$.

Apparently, the significance of condition (2.1) for the case of Hilbert space was first established in a paper of I. M. Gel′fand [3]. It is shown in

1) An operator is said to be a Volterra operator if it is completely continuous and quasi-nilpotent.

this paper that every chain of bounded vector variation in Hilbert space is similar to an orthogonal resolution of the identity.

We recall that a chain $\mathfrak{P} = \{P\}$ is said to be discrete if it has the form

$$\mathfrak{P} = \{0 = P_0 < P_1 < \cdots < P_n < \cdots < I\},$$

with $\dim \Delta P_j \mathfrak{B} = 1$ $(j = 1, 2, \cdots)$ and the sequence $\{P_n\}_0^\infty$ strongly converging to I. If the unit vector e_j is taken in each of the subspaces $P_j\mathfrak{B}$ then the system $\{e_j\}_1^\infty$ will form a basis of the space $\mathfrak{B}$.

It is easily verified that a discrete chain $\mathfrak{P}$ is a chain of bounded vector variation if and only if the system $\{e_j\}_1^\infty$ forms an unconditional basis of $\mathfrak{B}$.

We present some properties of chains of bounded vector variation which will be used below.

2°. *If $\mathfrak{P} = \{P\}$ is a chain of bounded vector variation then there exists a constant $C > 0$ such that*

$$\sup_{\mathfrak{Z};|\phi|\le 1} \sum_{j=1}^{n} |\phi(\Delta P_j x)| \le C|x| \quad (x \in \mathfrak{B}). \tag{2.2}$$

This proposition is easily proved with the use of a lemma of I. M. Gel'fand (see [4], Russian page 60).

Proposition 2° implies

3°. *The functional $\|x\|$ given by the equality*

$$\|x\| = \sup_{\mathfrak{Z};|\phi|\le 1} \sum_{j=1}^{n} |\phi(\Delta P_j x)| \quad (x \in \mathfrak{B}) \tag{2.3}$$

defines a norm in $\mathfrak{B}$ that is topologically equivalent to the original one.

In fact, by virtue of 2°, $\|x\| \le C|x|$ $(x \in \mathfrak{B})$. On the other hand, by considering the simplest partitioning $\mathfrak{Z} = \{0, I\}$ of $\mathfrak{P}$, we get from (2.3) that

$$\|x\| \ge \sup_{|\phi|\le 1} |\phi(x)| = |x|.$$

We note some elementary properties of the norm $\|\cdot\|$. It is easily verified that

$$\|x\| = \sup_{\mathfrak{Z}\supset\mathfrak{Z}_0;|\phi|\le 1} \sum_{j=1}^{n} |\phi(\Delta P_j x)| \quad (x \in \mathfrak{B}), \tag{2.4}$$

where $\mathfrak{Z}_0$ is any fixed partitioning of the chain $\mathfrak{P}$.

Formula (2.4) directly implies the propositions

4°. *If* $\mathfrak{Z} = \{P_j\}_0^n$ *is an arbitrary partitioning of a chain* $\mathfrak{P}$ *and* $\{\beta_j\}_1^n$ *are arbitrary numbers, then*

$$\left\| \sum_{j=1}^n \beta_j \Delta P_j x \right\| \leq \max_{1 \leq j \leq n} |\beta_j| \, \|x\| \quad (x \in \mathfrak{B}). \tag{2.5}$$

5°. *If* $P_1' \leq P_1 < P_2 \leq P_2'$ $(P_j, P_j' \in \mathfrak{P},\ j = 1, 2)$, *then*

$$\|(P_2 - P_1)x\| \leq \|(P_2' - P_1')x\| \quad (x \in \mathfrak{B}). \tag{2.6}$$

Hence, in particular, for any pair of projectors $P_1, P_2 \in \mathfrak{P}$ $(P_1 \neq P_2)$

$$\|P_1 - P_2\| = 1.$$

The following proposition holds.

6°. *Let* $\mathfrak{P} = \{P\}$ *be an arbitrary set of projectors from* $\mathfrak{R}$. *If the set* $\mathfrak{P}^* = \{P^*\}$ *of the adjoints of these projectors forms a chain of bounded vector variation then* $\mathfrak{P} = \{P\}$ *is also a chain of bounded vector variation.*

We first show that the chain $\mathfrak{P}$ is closed. Let $\{P_j\}_1^\infty$ be an arbitrary increasing sequence of projectors from $\mathfrak{P}$. Since $\mathfrak{P}^*$ is closed, the sequence $\{P_j^*\}_1^\infty$ strongly converges to a projector $P^* \in \mathfrak{P}^*$, and therefore $\sup \|P_j\| < \infty$.

We put $\mathfrak{M} = \bigcup_n P_n \mathfrak{B}$ and show that $\overline{\mathfrak{M}} = P\mathfrak{B}$. Clearly, $\overline{\mathfrak{M}} \subset P\mathfrak{B}$. Suppose there exists a vector $x \in P\mathfrak{B}$ such that $x \notin \overline{\mathfrak{M}}$. If ϕ is a functional from $\mathfrak{B}^*$ that is equal to zero on $\overline{\mathfrak{M}}$ and such that $\phi(x) = 1$ then $(P_j^*\phi)(x) = \phi(P_j x) = 0$ $(j = 1, 2, \cdots)$ and, inasmuch as $\lim P_j^* \phi = P^* \phi$, we get $(P^*\phi)(x) = 0$, i.e. $\phi(Px) = \phi(x) = 0$, which contradicts the fact that $\phi(x) = 1$.

Clearly, $\lim P_j x = x = Px$ for any $x \in \mathfrak{M}$. Since $\sup\|P_j\| < \infty$, we have $\lim P_j x = Px$ for all $x \in \overline{\mathfrak{M}} = P\mathfrak{B}$. Taking into account the fact that $P_j y = Py = 0$ for all $y \in (I - P)\mathfrak{B}$, we get that the sequence $\{P_j\}_1^\infty$ strongly converges to a projector $P \in \mathfrak{P}$.

We now show that $\mathfrak{P}$ is of bounded vector variation.

In fact, for any elements $\phi \in \mathfrak{B}^*$ and $\Phi \in \mathfrak{B}^{**}$

$$\sup_{\mathfrak{Z}} \sum_{j=1}^n |\Phi(\Delta P_j^* \phi)| < \infty.$$

If $\Phi(f) = f(x)$, where $x \in \mathfrak{B}$, then the latter condition becomes (2.1). The proposition is proved.

From here we arrive at the following assertion.

7°. *In a reflexive space a chain* $\mathfrak{P} = \{P\}$ *is a chain of bounded vector variation if and only if the chain* $\mathfrak{P}^* = \{P^*\}$ *is of bounded vector variation.*

2. We will assume below that the space $\mathfrak{B}$ is separable.

Let $\{x_j\}_1^\infty$ ($\|x_j\| = 1$, $j = 1, 2, \cdots$) be a dense set on the unit sphere of $\mathfrak{B}$.

If $P' < P''$ are arbitrary projectors of a chain $\mathfrak{P}$ of bounded vector variation then according to 5°

$$\|P' x_j\| \leq \|P'' x_j\| \quad (j = 1, 2, \cdots), \tag{2.7}$$

in which the strict inequality sign holds for at least one j by virtue of the density of the set $\{x_j\}_1^\infty$.

With each projector $P \in \mathfrak{P}$ we associate a real number $t = t(P)$ according to the rule

$$t = t(P) = \sum 2^{-j} \|P x_j\| \quad (P \in \mathfrak{P}). \tag{2.8}$$

Clearly $t(0) = 0$ and $t(I) = 1$. The range of the function $t = t(P)$ ($P \in \mathfrak{P}$) is denoted by Ω. To different projectors P' and P'' there correspond different numbers $t(P')$ and $t(P'')$; moreover, if $P' < P''$ then $t(P') < t(P'')$ by virtue of (2.7).

Thus the correspondence defined by formula (2.8) is one-to-one and we have the inverse function $P = P(t)$ ($t \in \Omega$), which is obviously a strictly increasing projector function defined on Ω whose range coincides with $\mathfrak{P}$.

We show that Ω is closed. Let t be an arbitrary limit point of Ω. Clearly t is the limit of a monotone sequence $\{t_j\}_1^\infty$ in Ω. The monotone sequence of projectors $P_j = P(t_j)$ ($\in \mathfrak{P}$, $j = 1, 2, \cdots$) has a projector $P_0 \in \mathfrak{P}$ as its limit by virtue of the fact that $\mathfrak{P}$ is closed. Therefore a point $t_0 \in \Omega$ can be found such that $P(t_0) = P_0$.

Clearly

$$|t_j - t_0| \leq \sum_{k=1}^{\infty} 2^{-k} \left| \|P_j x_k\| - \|P_0 x_k\| \right|. \tag{2.9}$$

Hence $t_0 = t = \lim t_j$. Consequently $t \in \Omega$ and $P(t) = P_0$.

It can also be shown that the projector function $P = P(t)$ is continuous on Ω.

We note further that if (α, β) is an interval adjoining Ω (at both α and β) then (P_α, P_β) $(P_\alpha = P(\alpha),\ P_\beta = P(\beta))$ is a gap of $\mathfrak{P}$, and, conversely, to each gap of $\mathfrak{P}$ there corresponds an interval adjoining Ω.[1] A chain $\mathfrak{P}$ is continuous if and only if the set Ω coincides with the interval $[0, 1]$. The projector function $P(t)$ is called a parametrization of $\mathfrak{P}$.

Thus every chain $\mathfrak{P}$ of bounded vector variation admits a parametrization by means of the corresponding closed set Ω.

If an operator function $F(P)$ with values in $\mathfrak{R}$ is strongly continuous on a chain $\mathfrak{P}$ of bounded vector variation, then clearly the vector function $F(P(t))x$ is uniformly continuous on the set Ω for any $x \in \mathfrak{B}$.

By using the uniform continuity of the vector function $F(P(t))x$ $(x \in \mathfrak{B})$ on Ω and the compactness of the set $\{Tx\colon\ x \in \mathfrak{B},\ \|x\| \le 1\}$, one can easily prove a more general proposition.

Lemma 1. *If an operator function $F(P)$ with values in $\mathfrak{R}$ is strongly continuous on a chain $\mathfrak{P}$ of bounded vector variation, then for any operator $T \in \mathfrak{S}_\infty$ the function $F(P(t))T$ is uniformly continuous on the set Ω in the sense of the operator norm.*

We will need another notion below.

A chain $\mathfrak{P}$ is called a chain of bounded operator variation if there exists a constant $k > 0$ such that for an arbitrary partitioning $\mathfrak{Z} = \{P_j\}_0^n$ of $\mathfrak{P}$ and for any operators $A_j \in \mathfrak{R}$ $(j = 1, 2, \cdots, n)$

$$\left|\sum_{j=1}^{n} \Delta P_j A_j \Delta P_j\right| \le k \max_{1\le j\le n} |\Delta P_j A_j \Delta P_j|. \tag{2.10}$$

The following proposition holds.

8°. *Every chain of bounded operator variation is a chain of bounded vector variation.*[2]

In fact, if we put $A_j = \epsilon_j I$, where $|\epsilon_j| = 1$ $(j = 1, 2, \cdots, n)$, we obtain from (2.10)

$$\left|\sum_{j=1}^{n} \epsilon_j \Delta P_j\right| \le k \max_{1\le j\le n} |\Delta P_j| \le C. \tag{2.11}$$

1) Hence a chain of bounded vector variation can have no more than a countable set of gaps.

2) The authors are indebted to A. S. Markus for this proposition.

If x is an arbitrary vector in $\mathfrak{B}$ and ϕ is an arbitrary functional in $\mathfrak{B}^*$, then by virtue of (2.11)

$$\left|\sum_{j=1}^{n} \epsilon_j \phi(\Delta P_j x)\right| \leq C|\phi||x|. \tag{2.12}$$

Putting $\epsilon_j = \exp(-i \arg \phi(\Delta P_j x))$, we obtain

$$\sum_{j=1}^{n} |\phi(\Delta P_j x)| \leq C|\phi||x|, \tag{2.13}$$

which directly implies condition (2.1).

The proposition is proved.

The converse proposition also holds for Hilbert space. This follows from the above-mentioned theorem of I. M. Gel'fand [3], according to which every chain of bounded vector variation in Hilbert space is similar to an orthogonal resolution of the identity.

Clearly, the chain $\mathfrak{P} = \{P_\alpha\}$ defined in L_p $(1 \leq p \leq \infty)$ by the equality

$$(P_\alpha x)(t) = \begin{cases} x(t), & 0 \leq t \leq \alpha \\ 0, & \alpha \leq t \leq 1 \end{cases},$$

is a chain of bounded vector and of bounded operator variation.

§ 3. Convergence of some integrals over a chain

1. An operator $T \in \mathfrak{R}$ is said to be nuclear if it is representable in the form

$$Tx = \sum_{r=1}^{\infty} a_r \phi_r(x) \eta_r \quad (x \in \mathfrak{B}), \tag{3.1}$$

where the a_r are complex numbers satisfying the condition

$$\sum_{r=1}^{\infty} |a_r| < \infty, \tag{3.2}$$

while $\phi_r \in \mathfrak{B}^*$, $\eta_r \in \mathfrak{B}$ and $|\phi_r| = |\eta_r| = 1$ $(r = 1, 2, \cdots)$.

With each nuclear operator T there is associated its nuclear norm

$$|T|_1 = \inf \sum_{r=1}^{\infty} |a_r|,$$

where the infimum is taken over all of the representations of T in the form (3.1).

The set of all nuclear operators in $\mathfrak{R}$ is denoted by $\mathfrak{S}_1$.

Lemma 2. *Let $T \in \mathfrak{S}_1$, and let $\mathfrak{P}$ be an arbitrary chain of bounded vector variation. If an operator function $F(P)$ $(P \in \mathfrak{P})$ is bounded, then the integrals*

$$\int_{\mathfrak{P}}^{m} dPTF(P)dP, \quad \int_{\mathfrak{P}}^{M} dPTF(P)dP \tag{3.3}$$

strongly converge, and the following equalities hold:

$$\int_{\mathfrak{P}}^{m} dPTF(P)dP = \sum_{j=1}^{\omega} (P_j^+ - P_j^-)TF(P_j^-)(P_j^+ - P_j^-), \tag{3.4}$$

$$\int_{\mathfrak{P}}^{M} dPTF(P)dP = \sum_{j=1}^{\omega} (P_j^+ - P_j^-)TF(P_j^+)(P_j^+ - P_j^-), \tag{3.5}$$

where $\{(P_j^-, P_j^+)\}_1^{\omega}$ is the sequence of all gaps of $\mathfrak{P}$.

In particular, if the chain $\mathfrak{P}$ is continuous then

$$\int_{\mathfrak{P}}^{m} dPTF(P)dP = \int_{\mathfrak{P}}^{M} dPTF(P)dP = 0.$$

Proof. Let ϵ be an arbitrary positive number and x any vector of $\mathfrak{B}$. There exists a natural number N such that for $j > N$

$$\|(P_j^+ - P_j^-)x\| < \epsilon \ (j > N, \ x \in \mathfrak{B}). \tag{3.6}$$

Let $\mathfrak{Z}_\epsilon = \{P_j\}_0^n$ denote a partitioning of $\mathfrak{P}$ containing the projectors $P_j^{\pm}$ $(j = 1, 2, \cdots, N)$ and having the property that any pair of projectors P_j, $P_{j+1} \in \mathfrak{Z}_\epsilon$ either coincides with a gap (P_k^-, P_k^+) $(k = 1, 2, \cdots, N)$ of $\mathfrak{P}$ or satisfies the inequality

$$\|\Delta P_j x\| < \epsilon. \tag{3.7}$$

By virtue of (3.1) we have

$$\left\| \sum_{j=N+1}^{N+P} (P_j^+ - P_j^-)TF(P_j^-)(P_j^+ - P_j^-)x \right\| \leq$$

$$\leq \sum_{r=1}^{\infty} |a_r| \left\| \sum_{j=N+1}^{N+P} \phi_r(F(P_j^-)(P_j^+ - P_j^-)x)(P_j^+ - P_j^-)\eta_r \right\|$$

and, according to 4°,

$$\left\| \sum_{j=N+1}^{N+P} (P_j^+ - P_j^-)TF(P_j^-)(P_j^+ - P_j^-)x \right\|$$

$$\leq \sum_{r=1}^{\infty} |a_r| \sup_j |\phi_r(F(P_j^-)(P_j^+ - P_j^-)x| \, \|\eta_r\|.$$

This inequality implies that

$$\left\| \sum_{j=N+1}^{N+P} (P_j^+ - P_j^-)TF(P_j^-)(P_j^+ - P_j^-)x \right\| \leq C \, \|F\| \sum_{r=1}^{\infty} |a_r| \epsilon, \tag{3.8}$$

where C is the constant of condition (2.2), and $\|F\| = \sup \|F(P)\|$ $(P \in \mathfrak{P})$. Hence the series (3.4) strongly converges.

The strong convergence of the series (3.5) is proved analogously.

Let the sums of these series be denoted by T_F^- and T_F^+ respectively.

Then for the partitioning $\mathfrak{Z}_\epsilon$, taking (3.8) into account, we obtain

$$\left\| \left[\sum_{j=1}^{n} \Delta P_j TF(P_{j-1}) \Delta P_j - T_F^- \right] x \right\|$$

$$\leq \left\| \sum_{j=1}^{n}{}' \Delta P_j TF(P_{j-1}) \Delta P_j x \right\| + C \|F\| \sum_{r=1}^{\infty} |a_r| \epsilon,$$

where the prime means that the summation is taken over those values of j for which ΔP_j does not coincide with one of the operators $P_k^+ - P_k^-$ $(k = 1, 2, \cdots, N)$.

Therefore, according to (3.7),

$$\left\| \sum_{j=1}^{n}{}' \Delta P_j TF(P_{j-1}) \Delta P_j x \right\| \leq C \|F\| \sum_{r=1}^{\infty} |a_r| \epsilon$$

and hence

$$\left\| \left[\sum_{j=1}^{n} \Delta P_j TF(P_{j-1}) \Delta P_j - T_F^- \right] x \right\| \leq 2C \|F\| \sum_{r=1}^{\infty} |a_r| \epsilon. \tag{3.9}$$

Since this inequality also holds for an arbitrary partitioning $\mathfrak{Z} \supset \mathfrak{Z}_\epsilon$, it implies the validity of equality (3.4).

Equality (3.5) is proved in exactly the same way.

The lemma is proved.

This lemma directly implies the following

Corollary. *If* $T \in \mathfrak{S}_1$ *and* $\mathfrak{P}$ *is a chain of bounded vector variation, then the integral*

$$\int_{\mathfrak{P}} dPTdP \tag{3.10}$$

strongly converges, and the following equality holds:

$$\int_{\mathfrak{P}} dPTdP = \sum_{j=1}^{\omega} (P_j^+ - P_j^-)T(P_j^+ - P_j^-).$$

Theorem 3. *Let* $T \in \mathfrak{S}_1$, *and let* $\mathfrak{P}$ *be an arbitrary chain of bounded vector variation. If the function* $F(P)$ *is strongly continuous on* $\mathfrak{P}$, *then the integrals*

$$\overset{m}{\int_{\mathfrak{P}}} F(P)TdP \text{ and } \overset{M}{\int_{\mathfrak{P}}} F(P)TdP \tag{3.11}$$

converge uniformly, while the integrals

$$\overset{m}{\int_{\mathfrak{P}}} dPTF(P) \text{ and } \overset{M}{\int_{\mathfrak{P}}} dPTF(P) \tag{3.12}$$

converge strongly, with

$$\max\left\{\left\|\overset{m}{\int_{\mathfrak{P}}} F(P)TdP\right\|, \left\|\overset{M}{\int_{\mathfrak{P}}} F(P)TdP\right\|\right\} \leq C\|F\||T|_1 \tag{3.13}$$

and

$$\max\left\{\left\|\overset{m}{\int_{\mathfrak{P}}} dPTF(P)\right\|, \left\|\overset{M}{\int_{\mathfrak{P}}} dPTF(P)\right\|\right\} \leq C\,\|F\||T|_1. \tag{3.14}$$

Proof. Let $\{(P_j^-, P_j^+)\}_1^\omega$ be the sequence of all gaps of $\mathfrak{P}$, and ϵ an arbitrary positive number. We choose a natural number m so that

$$\sum_{r=m+1}^{\infty} |a_r| < \epsilon. \tag{3.15}$$

By virtue of the uniform continuity of the vector function $F(P(t))x$ $(x \in \mathfrak{B})$ on the set Ω there exists a natural number N and a partitioning $\mathfrak{Z}_\epsilon$ of $\mathfrak{P}$ containing the projectors P_j^+ $(j = 1, 2, \cdots, N)$ and having the property that any pair of projectors Q_j' and Q_j'' of $\mathfrak{P}$ lying between successive projectors of $\mathfrak{Z}_\epsilon$ either coincides with one of the gaps (P_k^-, P_k^+) $(k = 1, 2, \cdots, N)$ of $\mathfrak{P}$ or satisfies the inequality

$$\|(F(Q_j') - F(Q_j''))\eta_r\| < \epsilon \ (r = 1, 2, \cdots, m). \tag{3.16}$$

Let $\mathfrak{Z}_1$ and $\mathfrak{Z}_2$ be arbitrary extensions of $\mathfrak{Z}_\epsilon$ and let S_1 and S_2 be the partial sums of the first of the integrals (3.11) corresponding to $\mathfrak{Z}_1$ and $\mathfrak{Z}_2$ respectively. Let $\mathfrak{Z} = \{P_j\}_0^n$ denote the union of $\mathfrak{Z}_1$ and $\mathfrak{Z}_2$. Then it is easily seen that the difference $\Delta Sx = (S_1 - S_2)x$ can be represented in the form

$$\Delta Sx = \sum_{j=1}^{n} (F(Q_j') - F(Q_j''))T\Delta P_j x, \tag{3.17}$$

where Q_j' and Q_j'' $(j = 1, 2, \cdots, n)$ are a pair of projectors of $\mathfrak{P}$ lying between adjacent projectors of $\mathfrak{Z}_\epsilon$. It is obvious that all of the summands in this sum which correspond to the gaps (P_j^-, P_j^+) $(j = 1, 2, \cdots, N)$ of $\mathfrak{P}$ are equal to zero.

By taking into account (3.1), equality (3.17) can be rewritten in the form

$$\Delta Sx = \sum_{j=1}^{n} \sum_{r=1}^{\infty} a_r \phi_r(\Delta P_j x)(F(Q_j') - F(Q_j''))\eta_r,$$

which implies

$$\|\Delta Sx\| \le \sum_{r=1}^{m} |a_r| \sum_{j=1}^{n} |\phi_r(\Delta P_j x)| \|(F(Q_j') - F(Q_j''))\eta_r\|$$

$$+ \sum_{r=m+1}^{\infty} |a_r| \sum_{j=1}^{n} |\phi_r(\Delta P_j x)| \, (\|F(Q_j')\| + \|F(Q_j'')\|) \|\eta_r\|.$$

It follows from this inequality, according to (3.15) and (3.16), that

$$\|\Delta Sx\| < \epsilon\left[\sum_{r=1}^{\infty} |a_r| + 2C\,\|F\|\right]\|x\|,$$

and hence the first of the integrals (3.11) converges uniformly.

The uniform convergence of the second integral of (3.11) is proved analogously. Let $S_m(\mathfrak{Z})$ denote the integral sum of the first of the integrals (3.11) corresponding to an arbitrary patitioning $\mathfrak{Z} = \{P_j\}_0^n$ of $\mathfrak{P}$.

Then

$$Sm(\mathfrak{Z})x = \sum_{j=1}^{n} F(P_{j-1})T\Delta P_j x,$$

or

$$Sm(\mathfrak{Z})x = \sum_{j=1}^{n}\sum_{r=1}^{\infty} a_r\phi_r(\Delta P_j x)F(P_j)\eta_r,$$

which directly implies that

$$\left\|\int_{\mathfrak{P}}^{m} F(P)TdP\right\| \le C\|F\|\|T|_1.$$

Analogously,

$$\left\|\int_{\mathfrak{P}}^{M} F(P)TdP\right\| \le C\|F\|\|T|_1.$$

We now consider the first of the integrals (3.12).

If S_1 and S_2 are the integral sums of it corresponding respectively to partitionings $\mathfrak{Z}_1$ and $\mathfrak{Z}_2$ of the chain $\mathfrak{P}$, and if $\mathfrak{Z} = \{P_j\}_0^n$ is the union of $\mathfrak{Z}_1$ and $\mathfrak{Z}_2$, then

$$\|\,\Delta Sx\| \le \sum |a_r| \left\|\sum \phi_r((F(Q_i') - F(Q_j''))x)\Delta P_j\eta_r\right\| \quad (x \in \mathfrak{B}).$$

From this inequality one can easily deduce the strong convergence of the first integral of (3.12).

The strong convergence of the second of the integrals (3.12) is proved analogously.

Relation (3.14) is established in the same way as (3.13).

The theorem is proved.

From this theorem we directly obtain a proposition.

9°. *If* $\mathfrak{P}$ *is a chain of bounded vector variation, then for any operator* $T \in \mathfrak{S}_1$ *the integrals*

$$\int_{\mathfrak{P}}^{m} PTdP \text{ and } \int_{\mathfrak{P}}^{M} PTdP$$

converge uniformly, while the integrals

$$\int_{\mathfrak{P}}^{m} dPTP \text{ and } \int_{\mathfrak{P}}^{M} dPTP$$

converge strongly, and the following equalities hold:

$$\int_{\mathfrak{P}}^{M} PTdP = \int_{\mathfrak{P}}^{m} PTdP + \sum_{j=1}^{\omega} (P_j^+ - P_j^-)T(P_j^+ - P_j^-)$$

and

$$\int_{\mathfrak{P}}^{M} dPTP = \int_{\mathfrak{P}}^{m} dPTP + \sum_{j=1}^{\omega} (P_j^+ - P_j^-)T(P_j^+ - P_j^-).$$

2. We introduce into consideration another notion of an integral over a chain of projectors.

Let $F(P)$ and $G(P)$ be arbitrary operator functions defined on a chain $\mathfrak{P}$ with values in $\mathfrak{R}$. With these functions we form a function $S(\mathfrak{Z})$ by putting

$$S(\mathfrak{Z}) = \sum_{j=1}^{n} F(Q_j')\Delta P_j G(Q_j') \tag{3.18}$$

for each partitioning $\mathfrak{Z} = \{P_j\}_0^n$ of $\mathfrak{P}$; here $\{Q_j\}_1^n$ is an arbitrary collection of projectors from $\mathfrak{P}$ such that $P_{j-1} \leq Q_j \leq P_j$ $(j = 1, 2, \cdots, n)$.

If the operator function $S(\mathfrak{Z})$ has an operator $A \in \mathfrak{R}$ as its limit, we will say that the integral

$$\int_{\mathfrak{P}} F(P)dPG(P) \tag{3.19}$$

uniformly converges (or, simply, converges) and is equal to A.

The multiplicative integrals

$$\overrightarrow{\int_{\mathfrak{P}}}(I + F(P)dPG(P)), \quad \overleftarrow{\int_{\mathfrak{P}}}(I + F(P)dPG(P)) \tag{3.20}$$

and the integral

$$\int_{\mathfrak{P}} dPF(P)dP \tag{3.21}$$

are defined as the limits of the corresponding functions

$$\overrightarrow{\prod}(\mathfrak{Z}) = \overrightarrow{\prod_{j=1}^{n}}(I + F(Q_j')\Delta P_j G(Q_j')),$$

$$\overleftarrow{\prod}(\mathfrak{Z}) = \overleftarrow{\prod_{j=1}^{n}}(I + F(Q_j')\Delta P_j G(Q_j')),$$

$$\mathfrak{S}(\mathfrak{Z}) = \sum_{j=1}^{n} \Delta P_j F(Q_j)\Delta P_j.$$

Remark. It is obvious that if the integral (3.19) converges then the corresponding integrals (1.15) and (1.25) converge, with

$$\int_{\mathfrak{P}} F(P)dPG(P) = \int_{\mathfrak{P}}^{m} F(P)dPG(P) = \int_{\mathfrak{P}}^{M} F(P)dPG(P) \tag{3.22}$$

and the following condition being satisfied:

$$F(P_j^+)(P_j^+ - P_j^-)G(P_j^+) - F(P_j^-)(P_j^+ - P_j^-)G(P_j^-) = 0 \quad (j = 1, 2, \cdots, \omega), \tag{3.23}$$

where $\{(P_j^-, P_j^+)\}_1^{\omega}$ is the sequence of all gaps of $\mathfrak{P}$.

Theorem 4. *Let $T \in \mathfrak{S}_1$, let $\mathfrak{P}$ be a chain of bounded vector variation and let $F(P)$ be an operator function that is strongly continuous on $\mathfrak{P}$. In order for the integral*

$$\int_{\mathfrak{P}} F(P)TdP \tag{3.24}$$

to converge uniformly it is necessary and sufficient that the following equalities be satisfied:

$$(F(P_j^+) - F(P_j^-))T(P_j^+ - P_j^-) = 0 \quad (j = 1, 2, \cdots, \omega), \tag{3.25}$$

where $\{(P_j^-, P_j^+)\}_1^\omega$ *is the sequence of all gaps of* $\mathfrak{P}$.

In particular, if $\mathfrak{P}$ *is continuous then the integral* (3.24) *always converges uniformly.*

Proof. The necessity of the conditions (3.25) follows from the Remark made above.

The sufficiency of these conditions follows from Theorem 3.

In fact, if the equalities (3.25) hold then the integrals (3.11), which converge uniformly by virtue of Theorem 3, are equal to each other. Repeating the corresponding arguments from the proof of this theorem, we get that the integral (3.24) converges uniformly.

The theorem is proved.

In the case of Hilbert space ($\mathfrak{B} = \mathfrak{H}$) this theorem was first proved by M. S. Brodskiĭ [5] under the assumption that $\mathfrak{P}$ is a chain of orthogonal projectors and $F(P) = P$. For the subsequent investigations of such integrals in Hilbert space see [1], [6] and [7].

An analogous theorem holds for the integral

$$\int_{\mathfrak{P}} dPTF(P). \tag{3.26}$$

In this case condition (3.25) is naturally replaced by the condition

$$(P_j^+ - P_j^-)T(F(P_j^+) - F(P_j^-)) = 0 \quad (j = 1, 2, \cdots, \omega). \tag{3.27}$$

Lemma 2 can be strengthened for the case of chains of bounded operator variation; namely, we have

Lemma 3. *Let* $T \in \mathfrak{S}_\infty$, *and let* $\mathfrak{P}$ *be an arbitrary chain of bounded operator variation. If an operator function* $F(P)$ *is bounded on* $\mathfrak{P}$, *then the integrals*

$$\overset{m}{\int_{\mathfrak{P}}} dPTF(P)dP \quad \text{and} \quad \overset{M}{\int_{\mathfrak{P}}} dPTF(P)dP \tag{3.28}$$

converge uniformly, and the following equalities hold:

$$\int_{\mathfrak{P}}^{m} dPTF(P)dP = \sum_{j=1}^{\omega}(P_j^+ - P_j^-)TF(P_j^-)(P_j^+ - P_j^-) \tag{3.29}$$

and

$$\int_{\mathfrak{P}}^{M} dPTF(P)dP = \sum_{j=1}^{\omega}(P_j^+ - P_j^-)TF(P_j^+)(P_j^+ - P_j^-), \tag{3.30}$$

where $\{(P_j^-, P_j^+)\}_1^\omega$ *is the sequence of all gaps of* $\mathfrak{P}$.

In particular, if $\mathfrak{P}$ *is continuous then*

$$\int_{\mathfrak{P}}^{m} dPTF(P)dP = \int_{\mathfrak{P}}^{M} dPTF(P)dP = \int_{\mathfrak{P}} dPTF(P)dP = 0. \tag{3.31}$$

Proof. Let ϵ be an arbitrary positive number. By virtue of Lemma 1 there exists a natural number N such that for $j > N$

$$\|(P_j^+ - P_j^-)T\| < \epsilon. \tag{3.32}$$

This result and (2.10) imply that the series (3.29) and (3.30) converge uniformly. We denote their sums by T_F^- and T_F^+ respectively. Clearly, T_F^- and T_F^+ belong to $\mathfrak{S}_\infty$.

Let $\mathfrak{Z}_\epsilon = \{P_j\}_0^n$ be a partitioning of $\mathfrak{P}$ containing all of the projectors $P_j^\pm$ $(j = 1, 2, \cdots, N)$ and having the property that any pair of projectors P_j, $P_{j+1} \in \mathfrak{Z}_\epsilon$ of which at least one is not among the $P_j^\pm$ $(j = 1, 2, \cdots, N)$ satisfies the inequality

$$\|\Delta P_j T\| < \epsilon. \tag{3.33}$$

Then, inasmuch as

$$\left\|\sum_{j=1}^{n} \Delta P_j TF(P_{j-1})\Delta P_j - T_F^-\right\|$$

$$\leq \left\|\sum_{j=1}^{n} \Delta P_j TF(P_{j-1})\Delta P_j - \sum_{j=1}^{\omega}(P_j^+ - P_j^-)TF(P_j^-)(P_j^+ - P_j^-)\right\|$$

$$+ \left\|\sum_{j=N+1}^{\omega} (P_j^+ - P_j^-)TF(P_j^-)(P_j^+ - P_j^-)\right\|,$$

we get by virtue of (2.10) and (3.32) that

$$\left\| \sum_{j=1}^{n} \Delta P_j T F(P_{j-1}) \Delta P_j - T_F^- \right\| \le k\|F\|(\max\|\Delta P_j T\| + \epsilon), \tag{3.34}$$

where the maximum is taken over all ΔP_j not coinciding with one of the operators $P_j^+ - P_j^-$ $(j = 1, 2, \cdots, N)$ and $\|F\| = \sup\|F(P)\|$ $(P \in \mathfrak{P})$.

From (3.34) and (3.33) we obtain equality (3.29).

Equality (3.30) is proved analogously.

The lemma is proved.

Corollary. *If* $T \in \mathfrak{S}_\infty$ *and* $\mathfrak{P}$ *is a chain of bounded operator variation, then the integral*

$$\int_{\mathfrak{P}} dPTdP \tag{3.35}$$

converges uniformly and the following equality holds:

$$\int_{\mathfrak{P}} dPTdP = \sum_{j=1}^{\omega} (P_j^+ - P_j^-)T(P_j^+ - P_j^-). \tag{3.36}$$

In particular, if $\mathfrak{P}$ *is continuous then the integral* (3.35) *is equal to zero.*

§ 4. Factorization of operators close to the identity operator

Theorem 5. *Suppose an operator* $A = I + T$ $(T \in \mathfrak{S}_\infty)$ *satisfies condition* a) *and the inequality*

$$\sup \|(I + PTD)^{-1}\| < \infty, \ (P \in \mathfrak{P}). \tag{4.1}$$

If $\mathfrak{P}^* = \{P^*\}$ *is a chain of bounded vector variation then* A *admits the factorization* (1.1) *relative to the chain* $\mathfrak{P} = \{P\}$, *and consequently formulas* (1.40)–(1.47) *hold for the factors of this factorization.*

Proof. By virtue of proposition 7°, the chain $\mathfrak{P} = \{P\}$ is also a chain of bounded vector variation, and therefore, according to Theorem 3, the integrals

$$\int_{\mathfrak{P}}^{m} (I + PTP)^{-1} PTdP \quad \text{and} \quad \int_{\mathfrak{P}}^{m} ((I + PTP)^{-1})^* P^* T^* dP^*$$

converge uniformly.

It is then easily seen that the integral

$$\int_{\mathfrak{P}}^{m} dPTP(I + PTP)^{-1},$$

also converges uniformly, and the proof is completed by making use of the Corollary of Theorem 2.

We note that equalities (1.42) and (1.45) for the diagonal factors D and D^{-1} can be written by virtue of Lemma 2 in the form

$$D = I + \sum_{j=1}^{\omega} (P_j^+ - P_j^-)T(I - P_j^-(I + P_j^- TP_j^-)^{-1}P_j^- T)(P_j^+ - P_j^-), \tag{4.2}$$

$$D^{-1} = I - \sum_{j=1}^{\omega} (P_j^+ - P_j^-)TP_j^+(I + P_j^+ TP_j^+)^{-1}(P_j^+ - P_j^-), \tag{4.3}$$

where $\{(P_j^-, P_j^+)\}_1^\omega$ is the sequence of all gaps of $\mathfrak{P}$, and in which the two series converge uniformly.

In particular, if $\mathfrak{P}$ is continuous then $D = D^{-1} = I$.

In a reflexive space $\mathfrak{B}$, by virtue of 7°, the condition in Theorem 5 that the chain $\mathfrak{P}^* = \{P^*\}$ be of bounded vector variation can be replaced by the condition that the chain $\mathfrak{P} = \{P\}$ be of bounded vector variation.

In the case of a chain $\mathfrak{P} = \{P\}$ of bounded operator variation Theorem 2 implies according to Lemma 3 that if an operator $A = I + T$ $(T \in \mathfrak{S}_\infty)$ satisfies condition a) and inequality (4.1), and if at least one of the integrals (1.40), (1.41), (1.43), (1.44), (1.46) or (1.47) converges, then it admits the factorization (1.1) relative to $\mathfrak{P} = \{P\}$, and consequently the factors of this factorization satisfy formulas (1.40)–(1.47), with the diagonal factors D and D^{-1} also satisfying formulas (4.2)–(4.3).

From this and from Theorem 3 we obtain

Theorem 6. *If $\mathfrak{P}$ is a chain of bounded operator variation then any operator $a = I + T$ $(T \in \mathfrak{S}_1)$ satisfying condition* a) *and inequality* (4.1) *admits the factorization* (1.1) *relative to $\mathfrak{P}$.*

Further, if an invertible operator $A = I + T$ $(T \in \mathfrak{S}_\infty)$ can be represented in the form $A = Y_- Y_+$, where $Y_\pm - I \in \mathfrak{S}_\infty$ and $Y_\pm \in \mathfrak{R}_\pm(\mathfrak{P})$, and $\mathfrak{P}$ is a chain of bounded operator variation, then A admits the factorization (1.1) *relative to $\mathfrak{P}$.*

BIBLIOGRAPHY

[1] I. C. Gohberg and M. G. Kreĭn, *On the factorization of operators in Hilbert space*, Acta Sci. Math. Szeged. 25 (1964), 90–123; English transl., Amer. Math. Soc. Transl. (2) 51 (1966), 155–188. MR 29 #6313.

[2] M. A. Barkar′ and I. C. Gohberg, *On the factorization of operators with respect to a discrete chain of projectors in a Banach space*, Mat. Issled. 1 (1966), no. 1, 32–54; English transl., Amer. Math. Soc. Transl. (2) 90 (1970), 81–103. MR 34 #6529.

[3] I. M. Gel′fand, *Remark on the work of N. K. Bari, "Biorthogonal systems and bases in Hilbert space"*, Moskov. Gos. Univ. Učen. Zap. 148, Mat. 4 (1951), 224–225. (Russian) MR 14, 289.

[4] N. I. Ahiezer and I. M. Glazman, *Theory of linear operators in Hilbert space*, GITTL, Moscow, 1950; English transl., Ungar, New York, 1961. MR 13, 358.

[5] M. S. Brodskiĭ, *On the triangular representation of completely continuous operators with one-point spectra*, Uspehi Mat. Nauk 16 (1961), no. 1 (97), 135–141; English transl., Amer. Math. Soc. Transl. (2) 47 (1965), 59–65. MR 24 #A426.

[6] M. S. Brodskiĭ, I. C. Gohberg, M. G. Kreĭn and V. I. Macaev, *Some investigations in the theory of non-selfadjoint operators*, Proc. Fourth All-Union Math. Congress, vol. II: Sectional Lectures, "Nauka," Leningrad, 1964, pp. 261–271; English transl., Amer. Math. Soc. Transl. (2) 65 (1967), 237–251. MR 36 #3153.

[7] I. C. Gohberg and M. G. Kreĭn, *Theory and applications of Volterra operators in Hilbert space*, "Nauka", Moscow, 1967; English transl., Transl. Math. Monographs, vol. 24, Amer. Math. Soc., Providence, R. I., 1970. MR 36 #2007.

Translated by:
S. Smith

Amer. Math. Soc. Transl.
(2) Vol. 90, 1970

ON ONE-PARAMETER FAMILIES OF EXTENSIONS OF A SYMMETRIC OPERATOR*

UDC 517.9

A. V. ŠTRAUS

We investigate a symmetric operator A and its selfadjoint extension $\tilde{A}$ with range in a space. To each complex number λ, including ∞, there corresponds an operator B_λ with domain of definition $\mathfrak{L}_\lambda$, which depends on λ; moreover $A \subset B_\lambda \subset A^*$. The linear manifold $\mathfrak{L}_\mu$ for arbitrary real μ and also for $\mu = \infty$ is described in terms of the boundary values of some analytic operator function.

Introduction

In his well-known investigations M. A. Naĭmark [1], [2] constructs a theory of selfadjoint extensions of a symmetric operator with range in a given Hilbert space, and he obtains some important results relative to the spectral theory of a symmetric operator. Some related questions are investigated in article [3] of the author, to which the present article is closely related. Let us recall some definitions and propositions from [3].

Let A be a closed symmetric nonselfadjoint operator in the Hilbert space H with domain of definition D_A, dense in H; let $\tilde{A}$ be a selfadjoint extension of A with range in the Hilbert space $\tilde{H} \supset H$. To each complex number λ there corresponds a linear manifold

$$\tilde{\mathfrak{L}}_\lambda = \{\tilde{f}:\ \tilde{A}\tilde{f} - \lambda\tilde{f} \in H\}.$$

We denote by P the orthogonal projection operator of $\tilde{H}$ onto H, and we introduce the one-parameter family of linear manifolds

$$\mathfrak{L}_\lambda = P\tilde{\mathfrak{L}}_\lambda.$$

For any λ

$$D_A \subset \mathfrak{L}_\lambda \subset D_{A^*},$$

where A^* is the operator adjoint to A.

We define the one-parameter family of extensions B_λ of the operator A, letting

$$B_\lambda = A^* \,|\, \mathfrak{L}_\lambda;$$

which means that B_λ is the restriction of the operator A^* to $\mathfrak{L}_\lambda$, so that for any complex λ

$$A \subset B_\lambda \subset A^*$$

* Translation of Izv. Akad. Nauk SSSR Ser. Mat. 30 (1966), 1325–1352.

We will say that the one-parameter families $\mathfrak{L}_\lambda$ and B_λ are generated by the selfadjoint extension $\widetilde{A}$ of A.

For any nonreal λ the operator $B_\lambda - \lambda E$ has a bounded inverse defined on all of H, and

$$(B_\lambda - \lambda E)^{-1} = P(\widetilde{A} - \lambda E)^{-1}|_H, \tag{0.1}$$

i.e. the operator function $(B_\lambda - \lambda E)^{-1}$ $(\mathrm{Im}\,\lambda \neq 0)$ coincides with the so-called generalized resolvent of A generated by $\widetilde{A}$.

Let $\mathfrak{N}_\lambda$ denote the defect subspace of A corresponding to the nonreal value λ, so that

$$\mathfrak{N}_\lambda = H \ominus (A - \lambda E) D_A = \{h\colon\ A^* h = \bar{\lambda} h\}.$$

We now fix an arbitrary nonreal number λ_0 and introduce the following notation for half-planes:

$$\Pi = \{\lambda\colon\ \mathrm{Im}\,\lambda \cdot \mathrm{Im}\,\lambda_0 > 0\}, \quad \Pi^* = \{\lambda\colon\ \mathrm{Im}\,\lambda \cdot \mathrm{Im}\,\lambda_0 < 0\}.$$

From the formula for generalized resolvents established earlier by the author in [4], it follows that

$$\mathfrak{L}_\lambda = \begin{cases} D_A + [F(\lambda) - E]\,\mathfrak{N}_{\lambda_0} & (\lambda \in \Pi), \\ D_A + [F^*(\bar{\lambda}) - E]\,\mathfrak{N}_{\bar{\lambda}_0} & (\lambda \in \Pi^*), \end{cases} \tag{0.2}$$

where $F(\lambda)$ is some operator function which is analytic in the half-plane Π and whose values are linear nonexpansive operators from $\mathfrak{N}_{\lambda_0}$ into $\mathfrak{N}_{\bar{\lambda}_0}$; conversely, if $F(\lambda)$ $(\lambda \in \Pi)$ is any operator function possessing the indicated properties, then the family of linear manifolds $\mathfrak{L}_\lambda$ $(\mathrm{Im}\,\lambda \neq 0)$ defined by formula (0.2) is generated by some selfadjoint extension $\widetilde{A}$ of A. We agree to say in this case that $F(\lambda)$ corresponds to the selfadjoint extension $\widetilde{A}$ of A.

The problem of describing the linear manifold $\mathfrak{L}_\mu$ for any real μ with the help of the operator function $F(\lambda)$ is of interest from the point of view of spectral theory. A considerable part of [3] is devoted to this question. For any real μ the operator B_μ is a symmetric extension of A (we do not exclude the case $B_\mu = A$) and therefore the linear manifold $\mathfrak{L}_\mu$ is representable in the form

$$\mathfrak{L}_\mu = D_A + (V_\mu - E)\,\mathfrak{N}_{\lambda_0,\,\mu}, \tag{0.3}$$

where $\mathfrak{N}_{\lambda_0,\,\mu}$ is some linear manifold in the defect subspace $\mathfrak{N}_{\lambda_0}$ and V_μ is some isometric operator with domain of definition $\mathfrak{N}_{\lambda_0,\,\mu}$ and range of values in $\mathfrak{N}_{\bar{\lambda}_0}$. In particular the following propositions appear in [3].

I. *For some sequence of numbers* $\lambda_n \in \Pi$ *converging to* μ *and for some sequence of elements* $\psi_n \in \mathfrak{N}_{\lambda_0}$ *strongly converging to the element* ψ, *let the double sequence of numbers*

$$\omega_{mn} = \frac{1}{\lambda_n - \overline{\lambda}_m} [(\psi_n, \psi_m) - (F(\lambda_n)\psi_n, F(\lambda_m)\psi_m)] \tag{0.4}$$

have a finite limit as m *and* n *simultaneously tend to* ∞. *Then* $\psi \in \mathfrak{N}_{\lambda_0, \mu}$ *and* $\lim_{n\to\infty} F(\lambda_n)\psi = V_\mu \psi$.*

For any fixed ϵ such that $0 < \epsilon < \pi/2$ we will assume

$$\Pi_{\mu,\varepsilon} = \{\lambda\colon \lambda \in \Pi, \quad \varepsilon < |\arg(\lambda - \mu)| < \pi - \varepsilon\}.$$

II. *Let* $\psi \in \mathfrak{N}_{\lambda_0, \mu}$. *Then for any sequence of numbers* $\lambda_n \in \Pi_{\mu,\epsilon}$ *converging to* μ, *there exists a sequence of elements* $\psi_n \in \mathfrak{N}_{\lambda_0}$, *strongly converging to* ψ, *such that the double sequence* ω_{mn} *defined by* (0.4) *has a finite limit as* m *and* n *simultaneously tend to* ∞.

From propositions I and II it follows in particular that for any $\psi \in \mathfrak{N}_{\lambda_0, \mu}$

$$\lim_{\substack{\lambda\to\mu \\ \lambda\in\Pi_{\mu,\varepsilon}}} F(\lambda)\psi = V_\mu \psi.$$

These propositions I and II together with (0.3) describe the linear manifold $\mathfrak{L}_\mu$ for any real μ.

III. *In order that the element* $\phi - \psi$, *where* $\phi \in \mathfrak{N}_{\overline{\lambda}_0}$, $\psi \in \mathfrak{N}_{\lambda_0}$, *belong to the linear manifold* $\mathfrak{L}_\mu$, *it is sufficient that the following conditions be satisfied*: 1) $\|\phi\| = \|\psi\|$; 2) *for some sequence of numbers* $\lambda_n \in \Pi_{\mu,\epsilon}$ *converging to* μ, *the sequence of elements*

$$\frac{1}{\lambda_n - \mu} [F(\lambda_n)\psi - \varphi] \tag{0.5}$$

*is weakly convergent.***

In frequent case when the operator A has defect index (1,1), i.e. when both defect subspaces $\mathfrak{N}_{\lambda_0}$ and $\mathfrak{N}_{\overline{\lambda}_0}$ are one-dimensional, the conditions 1) and 2) of proposition III are not only sufficient but also necessary in order that $\phi - \psi$ belong to the linear manifold $\mathfrak{L}_\mu$; moreover, if $\phi - \psi \in \mathfrak{L}_\mu$, then the sequence (0.5) is convergent whatever the sequence $\lambda_n \in \Pi_{\mu,\epsilon}$ converging to μ may be; that is, the vector-function $F(\lambda)\psi$ has an angular boundary

* The symbol lim as applied to sequences of elements of a Hilbert space and to vector-functions with values in this space is understood to mean limit in the sense of strong convergence. The symbol $\to$ is also used to indicate strong convergence.

** The corresponding theorem in [3] contains still another condition: $\lim_{n\to\infty} F(\lambda_n)\psi = \phi$. This condition is superfluous. In fact, if condition 2) is satisfied, then $F(\lambda_n)\psi$ converges weakly to ϕ; if in addition condition 1) is satisfied, then $\|F(\lambda_n)\psi\| \leq \|\phi\|$ and weak convergence implies strong convergence.

value ϕ and an angular derivative at the point μ. Thus the description of the linear manifold $\mathfrak{L}_\mu$ given in (3) can be formulated much more simply in the case of defect index (1, 1) than in the general case.

In the present article propositions I, II, and III are strengthened. With the help of the strengthened propositions we get a simple description of the linear manifolds $\mathfrak{N}_{\lambda_0, \mu}$ and $\mathfrak{L}_\mu$ in terms of the operator function $F(\lambda)$ for the general case of an operator A with arbitrary defect index. In particular we show that a necessary and sufficient condition that the element $\psi \in \mathfrak{N}_{\lambda_0}$ belong to the linear manifold $\mathfrak{N}_{\lambda_0, \mu}$ is the existence of the finite limit

$$\lim_{\substack{\lambda\to\mu \\ \lambda\in\Pi_{\mu,\,\varepsilon}}} \frac{\|\psi\| - \|F(\lambda)\psi\|}{\operatorname{Im}\lambda};$$

moreover, the following weaker condition is sufficient:

$$\sup_n \frac{\|\psi_n\| - \|F(\lambda_n)\psi_n\|}{|\operatorname{Im}\lambda_n|} < \infty$$

for some sequence of numbers $\lambda_n \in \Pi$ converging to μ and some sequence of elements $\psi_n \in \mathfrak{N}_{\lambda_0}$ converging weakly to ψ.

The study of the linear manifold $\mathfrak{L}_\mu$ is closely connected in the present article with the study of the linear manifold $D_{\widetilde{A}} \cap H$ in terms of the same operator function $F(\lambda)$. The relationship between the two questions is very close: if we observe that for $\lambda \neq 0$

$$\widetilde{\mathfrak{L}}_\lambda = \left\{\tilde{f}\colon \tilde{f} - \frac{1}{\lambda}\widetilde{A}\tilde{f} \in H\right\},$$

then it is natural to assume

$$\mathfrak{L}_\infty = \widetilde{\mathfrak{L}}_\infty = D_{\widetilde{A}} \cap H.$$

Using the results related to the linear manifold $\mathfrak{L}_\infty$ and the converse proposition related to formula (0.2), we get some theorems on the boundary properties of an arbitrary operator-function $\Phi(\lambda)$ $(\lambda \in \Pi)$ satisfying the same conditions as $F(\lambda)$. These propositions are perhaps of independent interest. A simple consequence of one of them is the theorem of Carathéodory about angular boundary derivatives of functions analytic and bounded in the unit circle. We use the boundary properties of the operator-function $F(\lambda)$ for the description of the linear manifolds $\mathfrak{L}_\mu$ and $\mathfrak{L}_\infty$.

Moreover in the present article we characterize the set of operator functions $F(\lambda)$ $(\lambda \in \Pi)$ corresponding to selfadjoint extensions $\widetilde{A} \supset A$ for which $D_{\widetilde{A}} \cap H = D_A$. Such operators $\widetilde{A}$, introduces and studied by M. A. Naĭmark [1], are called selfadjoint extensions of the second kind for A. It turns out that the operator-function $F(\lambda)$ corresponds to a selfadjoint extension of the second kind if and only if for any nonzero $\psi \in \mathfrak{N}_{\lambda_0}$

$$\lim_{\substack{\lambda\to\infty\\ \lambda\in\Pi_\varepsilon}} [|\lambda|(\|\psi\| - \|F(\lambda)\psi\|)] = +\infty,$$

where

$$\Pi_\varepsilon = \{\lambda:\ \lambda \in \Pi,\quad \varepsilon < |\arg \lambda| < \pi - \varepsilon\}.$$

The notations introduced here preserve their meaning throughout the present article except for certain individual cases which are treated in a special way. We now clarify a few more notational conventions used in this article:

$$\hat{H} = \tilde{H} \ominus H;$$

$\hat{P}$ is the orthogonal projection operator of $\tilde{H}$ onto $\hat{H}$; $P_{\mathfrak{N}}$ is the orthogonal projection operator of H onto the subspace $\mathfrak{N} \subset H$; the symbol $\rightharpoonup$ denotes weak convergence.

§1. Strengthening of propositions I and III

1. We will introduce some formulas which will be needed later on. If $\tilde{f} \in D_{\tilde{A}}$, then $P\tilde{f} \in D_{A^*}$ and $P\tilde{A}\tilde{f} = A^*P\tilde{f}$. Thus

$$\tilde{A}\tilde{f} = A^*P\tilde{f} + \hat{P}\tilde{A}\tilde{f}. \tag{1.1}$$

According to the definition of $\tilde{\mathfrak{L}}_\lambda$, for any complex number λ and any $\tilde{g} \in \tilde{\mathfrak{L}}_\lambda$ we have the equation

$$\hat{P}(\tilde{A} - \lambda E)\tilde{g} = 0;$$

that is,

$$\hat{P}\tilde{A}\tilde{g} = \lambda\hat{P}\tilde{g}. \tag{1.2}$$

By virtue of (1.1) and (1.2), for any $\tilde{g} \in \tilde{\mathfrak{L}}_\lambda$ we have

$$\tilde{A}\tilde{g} = A^*P\tilde{g} + \lambda\hat{P}\tilde{g}. \tag{1.3}$$

Let $\tilde{g}' \in \tilde{\mathfrak{L}}_{\lambda'}$, $\tilde{g}'' \in \tilde{\mathfrak{L}}_{\lambda''}$, where λ', λ'' are arbitrary complex numbers from the half-plane Π. According to (0.2)

$$P\tilde{g}' = f' + F(\lambda')\psi' - \psi', \qquad P\tilde{g}'' = f'' + F(\lambda'')\psi'' - \psi'',$$

where $f', f'' \in D_A$, and $\psi', \psi'' \in \mathfrak{N}_{\lambda_0}$. We have the following formula (cf. [3], Lemma 1.2):

$$(\tilde{g}', \tilde{g}'') = (P\tilde{g}', P\tilde{g}'') + \frac{\lambda_0 - \overline{\lambda_0}}{\lambda' - \overline{\lambda''}}[(\psi', \psi'') - (F(\lambda')\psi', F(\lambda'')\psi'')].$$

Since on the other hand

$$(\tilde{g}', \tilde{g}'') = (P\tilde{g}', P\tilde{g}'') + (\hat{P}\tilde{g}', \hat{P}\tilde{g}''),$$

we get the equation

$$(\hat{P}\widetilde{g}', \hat{P}\widetilde{g}'') = \frac{\lambda_0 - \overline{\lambda_0}}{\lambda' - \overline{\lambda''}} [(\psi', \psi'') - (F(\lambda')\psi', F(\lambda'')\psi'')]. \qquad (1.4)$$

Multiplying both sides by $\lambda'\overline{\lambda''}$ and taking (1.2) into account, we get

$$(\hat{P}\widetilde{A}\widetilde{g}', \hat{P}\widetilde{A}\widetilde{g}'') = (\lambda_0 - \overline{\lambda}_0)\frac{\lambda'\overline{\lambda}''}{\lambda' - \overline{\lambda}''} [(\psi', \psi'') - (F(\lambda')\psi', F(\lambda'')\psi'')]. \qquad (1.5)$$

2. We will establish here a formula which clarifies the meaning of the limit $\lim_{m,n\to\infty} \omega_{mn}$ mentioned in the hypothesis of proposition I.

So let $\{\psi_n\}$ be a sequence of elements from $\mathfrak{N}_{\lambda_0}$ strongly converging to ψ; let $\{\lambda_n\}$ be a sequence of a complex numbers from the half-plane Π converging to the real number μ; and assume the existence of a finite limit $\lim_{m,n\to\infty} \omega_{mn}$, where

$$\omega_{mn} = \frac{1}{\lambda_n - \overline{\lambda}_m} [(\psi_n, \psi_m) - (F(\lambda_n)\psi_n, F(\lambda_m)\psi_m)] \quad (m, n = 1, 2, \ldots).$$

For any n there exists an element $\widetilde{g}_n \in \widetilde{\mathfrak{L}}_{\lambda_n}$ such that

$$P\widetilde{g}_n = F(\lambda_n)\psi_n - \psi_n.$$

By virtue of (1.4)

$$(\hat{P}\widetilde{g}_n, \hat{P}\widetilde{g}_m) = (\lambda_0 - \overline{\lambda}_0)\,\omega_{mn} \quad (m, n = 1, 2, \ldots). \qquad (1.6)$$

As is shown in [3] (Theorem 2.2, which is the same as proposition I above), the sequence $\{\widetilde{g}_n\}$ strongly converges to an element $\widetilde{g} \in \widetilde{\mathfrak{L}}_\mu$ such that

$$P\widetilde{g} = \varphi - \psi,$$

where

$$\varphi = \lim F(\lambda_n)\psi_n,$$

and therefore $\psi \in \mathfrak{N}_{\lambda_0, \mu}$ and $V_\mu \psi = \phi$.

We will show that

$$\widetilde{g} \perp (\hat{H} \cap \widetilde{\mathfrak{L}}_\mu). \qquad (1.7)$$

First we note that

$$\hat{H} \cap \widetilde{\mathfrak{L}}_\mu = \hat{H} \cap \widetilde{\mathfrak{M}}_\mu, \qquad (1.8)$$

where

$$\widetilde{\mathfrak{M}}_\mu = \{\widetilde{f} : \widetilde{A}\widetilde{f} = \mu\widetilde{f}\}.$$

Actually, if $\widetilde{f} \in \hat{H} \cap \widetilde{\mathfrak{L}}_\mu$, then on the one hand

$$(\widetilde{A} - \mu E)\widetilde{f} \in \hat{H}, \qquad (1.9)$$

because for any $f \in D_A$ (since $\widetilde{f} \in \hat{H}$)

$$((\widetilde{A} - \mu E)\widetilde{f}, f) = (\widetilde{f}(, (A - \mu E)f) = 0,$$

and on the other

$$(\widetilde{A} - \mu E)\widetilde{f} \in H, \tag{1.10}$$

since $\widetilde{f} \in \widetilde{\mathfrak{L}}_{\mu}$. By virtue of (1.9) and (1.10)

$$(\widetilde{A} - \mu E)\widetilde{f} = 0,$$

that is, $\widetilde{f} \in \widetilde{\mathfrak{M}}_{\mu}$. So we have shown that

$$\hat{H} \cap \widetilde{\mathfrak{L}}_{\mu} \subset \widetilde{\mathfrak{M}}_{\mu},$$

and because $\widetilde{\mathfrak{M}}_{\mu} \subset \widetilde{\mathfrak{L}}_{\mu}$ equation (1.8) is valid.

For any $\widetilde{f} \in \hat{H} \cap \widetilde{\mathfrak{L}}_{\mu}$, taking (1.8) into account we have

$$(\widetilde{g}_n, \widetilde{f}) = \frac{1}{\mu - \lambda_n}(\widetilde{g}_n, (\widetilde{A} - \overline{\lambda}_n E)\ \widetilde{f}) = \frac{1}{\mu - \lambda_n}((\widetilde{A} - \lambda_n E)\widetilde{g}_n, \widetilde{f}) = 0, \tag{1.11}$$

and therefore also

$$\lim_{n\to\infty} (\widetilde{g}_n, \widetilde{f}) = 0;$$

that is,

$$(\widetilde{g}, \widetilde{f}) = 0,$$

and thus (1.7) is proved.

Thus the element $\widetilde{g}$ satisfies the conditions: 1) $\widetilde{g} \in \widetilde{\mathfrak{L}}_{\mu}$; 2) $P\widetilde{g} = V_{\mu}\psi - \psi$; 3) $\widetilde{g} \perp (\hat{H} \cap \widetilde{\mathfrak{L}}_{\mu})$. We will show that $\widetilde{g}$ is uniquely defined by these conditions. In fact let $\widetilde{g}'$ be any element satisfying the same conditions. Then $\widetilde{g} - \widetilde{g}' \in \widetilde{\mathfrak{L}}_{\mu}$, $\widetilde{g} - \widetilde{g}' \in \hat{H}$, and $\widetilde{g} - \widetilde{g}' \perp (\hat{H} \cap \widetilde{\mathfrak{L}}_{\mu})$, from which it follows that $\widetilde{g} - \widetilde{g}' = 0$.

Turning to equation (1.6) and taking into account that $\lim_{n\to\infty} \widetilde{g}_n = \widetilde{g}$, we arrive at the following formula, which supplements proposition I:

$$\lim_{m,\, n\to\infty} \omega_{mn} = \frac{1}{\lambda_0 - \overline{\lambda}_0} \| \hat{P}\widetilde{g} \|^2, \tag{1.12}$$

where $\widetilde{g}$ is the element uniquely defined by conditions 1), 2) and 3).

3. As we know, every bounded sequence of elements from a Hilbert space contains a weakly converging subsequence. From this property of weak compactness of bounded sets in Hilbert space the following elementary proposition, which will be needed later, follows immediately.

Lemma 1.1. *If a sequence of elements from a Hilbert space is bounded and all its weakly converging subsequences have the same weak limit, then the whole sequence converges weakly.*

Proof. Let $\{f_n\}$ be a sequence of elements from a Hilbert space X satisfying the indicated conditions and let f be the element which is the weak limit of each weakly converging subsequence of $\{f_n\}$. Now suppose that $\{f_n\}$ does not converge weakly to f. Then for some $\epsilon > 0$ and some $h \in X$ there is a subsequence $\{f_{n_k}\}$ such that

$$|(f_{n_k}-f,h)|\geqslant \varepsilon \quad (k=1,2,\ldots). \tag{1.13}$$

To simplify the notation let $g_k = f_{n_k}$ $(k = 1, 2, \cdots)$. Because $\{g_k\}$ is bounded it contains a weakly converging subsequence $\{g_{k_j}\}$. It is clear that $\{g_{k_j}\}$ is such a subsequence for $\{f_n\}$ and therefore $g_{k_j} \rightharpoonup f$ for $j \to \infty$. The latter is not possible because according to (1.13)

$$|(g_{k_j}-f,h)|\geqslant \varepsilon \quad (j=1,2,\ldots).$$

Thus $f_n \rightharpoonup f$ and the lemma is proved.

We now consider a stronger version of proposition I.

Theorem 1.1. *For some sequence of numbers $\lambda_n \in \Pi$ converging to the real number μ and for some sequence of elements $\psi_n \in \mathfrak{N}_{\lambda_0}$ weakly converging to the element ψ, let*

$$\sup_n \frac{\|\psi_n\| - \|F(\lambda_n)\psi_n\|}{|\operatorname{Im}\lambda_n|} < \infty. \tag{1.14}$$

Then $\psi \in \mathfrak{N}_{\lambda_0,\mu}$, the sequence $\{F(\lambda_n)\psi_n\}$ converges weakly to $V_\mu\psi$ and

$$\varliminf_{n\to\infty} \frac{\|\psi_n\|^2 - \|F(\lambda_n)\psi_n\|^2}{|\operatorname{Im}\lambda_n|} \geqslant \frac{1}{|\operatorname{Im}\lambda_0|}\|\hat{P}\tilde{g}\|^2, \tag{1.15}$$

where $\tilde{g}$ is an element uniquely defined by the following conditions: 1) $\tilde{g} \in \tilde{\mathfrak{L}}_\mu$; 2) $P\tilde{g} = V_\mu\psi - \psi$; 3) $\tilde{g} \perp (\hat{H} \cap \tilde{\mathfrak{L}}_\mu)$. *If the sequence $\{\psi_n\}$ converges strongly to ψ, then*

$$\lim_{n\to\infty} F(\lambda_n)\psi_n = \lim_{n\to\infty} F(\lambda_n)\psi = V_\mu\psi.$$

Proof. Because

$$F(\lambda_n)\psi_n - \psi_n \in \mathfrak{L}_{\lambda_n},$$

it follows that

$$F(\lambda_n)\psi_n - \psi_n = P\tilde{g}_n \quad (n = 1, 2, \ldots), \tag{1.16}$$

where $\tilde{g}_n \in \tilde{\mathfrak{L}}_{\lambda_n}$. By virtue of (1.4)

$$\|\hat{P}\tilde{g}_n\|^2 = \frac{\lambda_0 - \overline{\lambda_0}}{\lambda_n - \overline{\lambda}_n}(\|\psi_n\|^2 - \|F(\lambda_n)\psi_n\|^2). \tag{1.17}$$

Taking (1.14) and (1.17) into account we conclude that the sequence $\{\hat{P}\tilde{g}_n\}$ is bounded. Because $\{P\tilde{g}_n\}$ is also bounded, $\{\tilde{g}_n\}$ is bounded.

We consider an arbitrary weakly converging subsequence $\{\tilde{g}_{n_k}\}$ of the sequence $\{\tilde{g}_n\}$. Then for $k \to \infty$

$$\tilde{g}_{n_k} \rightharpoonup \tilde{g}, \tag{1.18}$$

where $\tilde{g}$ is some element from $\tilde{H}$. Hence it follows that

$$P\tilde{g}_{n_k} \rightharpoonup P\tilde{g};$$

that is,

$$F(\lambda_{n_k})\psi_{n_k} - \psi_{n_k} \rightharpoonup P\tilde{g}.$$

Therefore, assuming

$$\varphi = \psi + P\tilde{g}, \tag{1.19}$$

we have

$$F(\lambda_{n_k})\psi_{n_k} \rightharpoonup \varphi. \tag{1.20}$$

It is clear that $\phi \in \mathfrak{N}_{\overline{\lambda}_0}^{-}$.

According to (1.3)

$$\tilde{A}\tilde{g}_n = \lambda_0 F(\lambda_n)\psi_n - \overline{\lambda}_0\psi_n + \lambda_n \hat{P}\tilde{g}_n.$$

Taking (1.18) and (1.20) into account we conclude that for $k \to \infty$

$$\tilde{A}\tilde{g}_{n_k} \rightharpoonup \lambda_0\varphi - \overline{\lambda}_0\psi + \mu\hat{P}\tilde{g}. \tag{1.21}$$

As is easy to see, the graph of any selfadjoint operator is closed not only in the strong topology but also in a weak topology.* Therefore from (1.18) and (1.21) it follows that $\tilde{g} \in D_{\tilde{A}}$ and

$$\tilde{A}\tilde{g} = \lambda_0\varphi - \overline{\lambda}_0\psi + \mu\hat{P}\tilde{g}. \tag{1.22}$$

By virtue of the last equation

$$\tilde{A}\tilde{g} - \mu\tilde{g} \in H,$$

and this means that $\tilde{g} \in \tilde{\mathfrak{L}}_\mu$. Taking (1.19) into account, we now have

$$\varphi - \psi = P\tilde{g} \in \mathfrak{L}_\mu.$$

Hence it follows that $\psi \in \mathfrak{N}_{\lambda_0,\mu}$, $\phi = V_\mu\psi$ and

$$P\tilde{g} = V_\mu\psi - \psi. \tag{1.23}$$

According to (1.11), for any $\tilde{f} \in \hat{H} \cap \tilde{\mathfrak{L}}_\mu$

$$(\tilde{g}_n, \tilde{f}) = 0,$$

and therefore

$$(\tilde{g}, \tilde{f}) = \lim_{k\to\infty} (\tilde{g}_{n_k}, \tilde{f}) = 0;$$

that is,

$$\tilde{g} \perp (\hat{H} \cap \tilde{\mathfrak{L}}_\mu).$$

Thus $\tilde{g}$ satisfies conditions 1), 2), and 3) of the theorem. It was established in subsection 2 that $\tilde{g}$ is uniquely defined by these conditions. Therefore $\tilde{g}$ does not depend on the choice of the weakly converging subsequence $\{\tilde{g}_{n_k}\}$ of the bounded sequence $\{\tilde{g}_n\}$. Using Lemma 1.1, we conclude that for $n \to \infty$

$$\tilde{g}_n \rightharpoonup \tilde{g},$$

---

* An analogous proposition is true for the graph of an operator adjoint to an arbitrary linear operator.

and therefore

$$P\tilde{g}_n \rightharpoonup P\tilde{g} \tag{1.24}$$

and

$$\hat{P}\tilde{g}_n \rightharpoonup \hat{P}\tilde{g}. \tag{1.25}$$

By virtue of (1.16) and (1.23) we can rewrite (1.24) in the form

$$F(\lambda_n)\psi_n - \psi_n \rightharpoonup V_\mu\psi - \psi.$$

Because $\psi_n \rightharpoonup \psi$, it follows that

$$F(\lambda_n)\psi_n \rightharpoonup V_\mu\psi. \tag{1.26}$$

By virtue of a well-known estimate of the norm of a weak limit we get from (1.25) the inequality

$$\underline{\lim}_{n\to\infty} \|\hat{P}\tilde{g}_n\| \geqslant \|\hat{P}\tilde{g}\|. \tag{1.27}$$

Taking (1.17) and (1.27) into account, we get inequality (1.15).

We now assume that $\{\psi_n\}$ converges strongly to ψ. Then

$$\overline{\lim}_{n\to\infty} \|F(\lambda_n)\psi_n\| \leqslant \lim_{n\to\infty} \|\psi_n\| = \|\psi\| = \|V_\mu\psi\|,$$

which in conjunction with (1.26) proves strong convergence of $\{F(\lambda_n)\psi_n\}$ to $V_\mu\psi$. Because

$$\|F(\lambda_n)(\psi - \psi_n)\| \leqslant \|\psi - \psi_n\| \quad (n = 1, 2, \ldots),$$

the sequence $\{F(\lambda_n)\psi\}$ also converges strongly to $V_\mu\psi$.

The theorem is fully proved.

Observation 1. With the additional assumption that $\lambda_n \in \Pi_{\mu,\epsilon}$ $(n = 1, 2, \cdots)$, condition (1.14) is equivalent to the following:

$$\sup_n \frac{\|\psi_n\| - \|F(\lambda_n)\psi_n\|}{|\lambda_n - \mu|} < \infty.$$

In fact, for any $\lambda \in \Pi_{\mu,\epsilon}$

$$\frac{1}{|\lambda - \mu|} \leqslant \frac{1}{|\operatorname{Im}\lambda|} < \frac{1}{|\lambda - \mu|\sin\varepsilon}.$$

Observation 2. If for some sequence of numbers $\lambda_n \in \Pi$ converging to μ and for some sequence of elements $\psi_n \in \mathfrak{N}_{\lambda_0}$ weakly converging to ψ,

$$\lim_{n\to\infty} \frac{\|\psi_n\| - \|F(\lambda_n)\psi_n\|}{\operatorname{Im}\lambda_n} = 0, \tag{1.28}$$

then $V_\mu\psi - \psi \in D_{\tilde{A}}$, $\tilde{A}(V_\mu\psi - \psi) = \lambda_0 V_\mu\psi - \bar{\lambda}_0\psi$ and $F(\lambda)\psi = V_\mu\psi$ for any $\lambda \in \Pi$.

In fact, from (1.28) and (1.15) it follows that $P\tilde{g} = 0$, and therefore

$$\tilde{g} = P\tilde{g} = V_\mu\psi - \psi.$$

Taking (1.22) into account we get

$$\tilde{A}g = \lambda_0 V_\mu \psi - \overline{\lambda}_0 \psi.$$

Because $V_\mu \psi - \psi \in D_{\tilde{A}}$ and $\tilde{A}(V_\mu \psi - \psi) \in H$, $F(\lambda)\psi = V_\mu \psi$ for any $\lambda \in \Pi$ (cf. [3], §1. 3).

4. The following theorem is a strong version of proposition III.

Theorem 1.2 *Let* $\phi \in \mathfrak{N}_{\overline{\lambda}_0}$ *and* $\psi \in \mathfrak{N}_{\lambda_0}$ *satisfy the following conditions:* 1) $\|\phi\| = \|\psi\|$; 2) *for some sequence of numbers* $\lambda_n \in \Pi$ *converging to* μ

$$\sup_n \frac{|(F(\lambda_n)\psi - \varphi, \varphi)|}{|\operatorname{Im} \lambda_n|} < \infty. \tag{1.29}$$

Then $\phi - \psi \in \mathfrak{L}_\mu$ *and*

$$\lim_{n\to\infty} F(\lambda_n)\psi = \varphi. \tag{1.30}$$

Proof. By virtue of (1.29)

$$\lim_{n\to\infty} (F(\lambda_n)\psi - \varphi, \varphi) = 0. \tag{1.31}$$

Taking into account that

$$\|F(\lambda_n)\psi\| \leqslant \|\psi\| = \|\varphi\|,$$

we have

$$\|F(\lambda_n)\psi - \varphi\|^2 \leqslant 2[\|\varphi\|^2 - \operatorname{Re}(F(\lambda_n)\psi, \varphi)] = 2\operatorname{Re}(\varphi - F(\lambda_n)\psi, \varphi). \tag{1.32}$$

By virtue of (1.31) and (1.32)

$$\lim_{n\to\infty} \|F(\lambda_n)\psi - \varphi\| = 0,$$

so that equation (1.30) is proved.

For any elements $h', h'' \in H$ we have the equation

$$\|h'\|^2 - \|h''\|^2 \leqslant 2\operatorname{Re}(h' - h'', h'), \tag{1.33}$$

which is a consequence of the obvious formula

$$\|h'\|^2 - \|h''\|^2 = 2\operatorname{Re}(h' - h'', h') - \|h' - h''\|^2.$$

Letting $h' = \phi$ and $h'' = F(\lambda_n)\psi$ $(n = 1, 2, \cdots)$ in (1.33), and taking conditions 1) and 2) into account, we conclude that

$$\sup_n \frac{\|\psi\|^2 - \|F(\lambda_n)\psi\|^2}{|\operatorname{Im} \lambda_n|} < \infty,$$

and this means that

$$\sup_n \frac{\|\psi\| - \|F(\lambda_n)\psi\|}{\|\operatorname{Im} \lambda_n|} < \infty.$$

From Theorem 1.1 it now follows that $\psi \in \mathfrak{N}_{\lambda_0, \mu}$ and

$$\lim_{n\to\infty} F(\lambda_n)\psi = V_\mu \psi.$$

Combining the last formula with equation (1.30), we see that $\phi = V_\mu \psi$ and therefore $\phi - \psi \in \mathfrak{L}_\mu$. The theorem is thus proved.

Observation 1. If $\lambda_n \in \Pi_{\mu, \epsilon}$ $(n = 1, 2, \cdots)$, then condition (1.29) is equivalent to the following:

$$\sup_n \frac{|(F(\lambda_n)\psi - \varphi, \varphi)|}{|\lambda_n - \mu|} < \infty.$$

Observation 2. As is clear from the proof of the theorem, condition 2) can be replaced by the following weaker condition: for some sequence of numbers $\lambda_n \in \Pi$ converging to μ

$$\sup_n \frac{\operatorname{Re}(\varphi - F(\lambda_n)\psi, \varphi)}{|\operatorname{Im} \lambda_n|} < \infty.$$

§2. The linear manifold $\mathfrak{L}_\infty$

1. As we remarked earlier, it is natural to introduce the notation

$$\mathfrak{L}_\infty = D_{\widetilde{A}} \cap H.$$

The operator

$$B_\infty = P\widetilde{A}|_{\mathfrak{L}_\infty}$$

is a symmetric extension of A because for any $f, g \in \mathfrak{L}_\infty$

$$(B_\infty f, g) = (P\widetilde{A}f, g) = (\widetilde{A}f, g) = (f, \widetilde{A}g) = (f, P\widetilde{A}g) = (f, B_\infty g).$$

Obviously $\mathfrak{L}_\infty \subset D_{A^*}$ and

$$B_\infty = A^*|_{\mathfrak{L}_\infty}.$$

Since the linear manifold $\mathfrak{L}_\infty$ serves as the domain of definition of a symmetric extension of the operator A, it is representable in the form

$$\mathfrak{L}_\infty = D_A + (V_\infty - E)\mathfrak{N}_{\lambda_0, \infty}, \tag{2.1}$$

where $\mathfrak{N}_{\lambda_0, \infty}$ is some linear manifold in $\mathfrak{N}_{\lambda_0}$ and V_∞ is some isometric operator with domain of definition $\mathfrak{N}_{\lambda_0, \infty}$ and with range of values in $\mathfrak{N}_{\overline{\lambda}_0}$.*

We first consider a theorem analogous to proposition I as supplemented by formula (1.12).

Theorem 2.1. *For some sequence of numbers $\lambda_n \in \Pi$ converging to ∞ and for some sequence of elements $\psi_n \in \mathfrak{N}_{\lambda_0}$ converging strongly to the element*

* If both defect subspaces $\mathfrak{N}_{\lambda_0}$ and $\mathfrak{N}_{\overline{\lambda}_0}$ are infinite dimensional, then the operator B_∞ is not necessarily closed, i.e. the linear manifold $\mathfrak{N}_{\lambda_0, \infty}$ may not be a subspace.

ψ, let the double sequence of numbers

$$\gamma_{mn} = \frac{\lambda_n \overline{\lambda}_m}{\lambda_n - \overline{\lambda}_m} [(\psi_n, \psi_m) - (F(\lambda_n)\psi_n, F(\lambda_m)\psi_m)] \tag{2.2}$$

have a finite limit as m and n simultaneously tend to ∞. Then

$$\psi \in \mathfrak{N}_{\lambda_0, \infty}, \qquad \lim_{n\to\infty} F(\lambda_n)\psi_n = \lim_{n\to\infty} F(\lambda_n)\psi = V_\infty \psi$$

and

$$\lim_{m,\, n\to\infty} \gamma_{mn} = \frac{1}{\lambda_0 - \overline{\lambda}_0} \|\hat{P}\widetilde{A}(V_\infty\psi - \psi)\|^2.$$

Proof. First of all we note that by virtue of the conditions of the theorem

$$\lim_{m,\, n\to\infty} (F(\lambda_n)\psi_n, F(\lambda_m)\psi_m) = (\psi, \psi).$$

Hence the sequence $\{F(\lambda_n)\psi_n\}$ is fundamental, and thus it converges. We assume

$$\varphi = \lim_{n\to\infty} F(\lambda_n)\psi_n. \tag{2.3}$$

Obviously $\phi \in \mathfrak{N}_{\overline{\lambda}_0}$. Considering that $\psi_n \to \psi$ and $\|F(\lambda)\| \le 1$ for any $\lambda \in \Pi$, we conclude that

$$\lim_{n\to\infty} F(\lambda_n)\psi = \varphi. \tag{2.4}$$

Consider now a sequence of elements $\widetilde{g}_n \in \widetilde{\mathfrak{L}}_{\lambda_n}$ such that

$$P\widetilde{g}_n = F(\lambda_n)\psi_n - \psi_n \quad (n = 1, 2, \ldots). \tag{2.5}$$

Obviously

$$\widetilde{g}_n = F(\lambda_n)\psi_n - \psi_n + \hat{P}\widetilde{g}_n. \tag{2.6}$$

By virtue of (1.4) and (2.2)

$$\|\hat{P}\widetilde{g}_n\|^2 = \frac{\lambda_0 - \overline{\lambda}_0}{|\lambda_n|^2} \gamma_{nn} \quad (n = 1, 2, \ldots).$$

Hence it follows that

$$\lim_{n\to\infty} \hat{P}\widetilde{g}_n = 0.$$

Turning to equation (2.6) we conclude that the sequence $\{\widetilde{g}_n\}$ converges; moreover

$$\lim_{n\to\infty} \widetilde{g}_n = \varphi - \psi. \tag{2.7}$$

We investigate now the sequence $\{\widetilde{A}\widetilde{g}_n\}$. By virtue of (1.1) and (2.5)

$$\widetilde{A}\widetilde{g}_n = \lambda_0 F(\lambda_n)\psi_n - \overline{\lambda}_0\psi_n + \hat{P}\widetilde{A}\widetilde{g}_n \quad (n = 1, 2, \ldots). \tag{2.8}$$

According to (1.5) and (2.2)

$$(\hat{P}\widetilde{A}\widetilde{g}_n, \hat{P}\widetilde{A}\widetilde{g}_m) = (\lambda_0 - \overline{\lambda}_0)\gamma_{mn} \quad (m, n = 1, 2, \ldots). \tag{2.9}$$

Since the double sequence $\{\gamma_{mn}\}$ has a finite limit, the sequence $\{\hat{P}\widetilde{A}\widetilde{g}_n\}$ is fundamental because of (2.9), and therefore convergent since $\hat{H}$ is complete. By (2.8) we can now conclude that $\{\widetilde{A}\widetilde{g}_n\}$ is convergent and

$$\lim_{n\to\infty} \widetilde{A}\widetilde{g}_n = \lambda_0\varphi - \overline{\lambda}_0\psi + \hat{h}, \tag{2.10}$$

where

$$\hat{h} = \lim_{n\to\infty} \hat{P}\widetilde{A}\widetilde{g}_n.$$

We observe that by (2.9)

$$\lim_{m,\, n\to\infty} \gamma_{mn} = \frac{1}{\lambda_0 - \overline{\lambda}_0}\|\hat{h}\|^2. \tag{2.11}$$

From (2.7) and (2.10) it follows that

$$\varphi - \psi \in D_{\widetilde{A}}$$

and

$$\widetilde{A}(\varphi - \psi) = \lambda_0\varphi - \overline{\lambda}_0\psi + \hat{h}.$$

Therefore

$$\psi \in \mathfrak{N}_{\lambda_0,\,\infty}, \quad \varphi = V_\infty\psi \tag{2.12}$$

and

$$\hat{h} = \hat{P}\widetilde{A}(V_\infty\psi - \psi). \tag{2.13}$$

Comparing formulas (2.3), (2.4), and (2.12), and also (2.11) and (2.13), we see that the theorem is true.

The following proposition is an analog of Theorem 1.1.

Theorem 2.2. *For some sequence of numbers* $\lambda_n \in \Pi$ *converging to* ∞ *and for some sequence of elements* $\psi_n \in \mathfrak{N}_{\lambda_0}$ *weakly converging to the element* ψ, *let*

$$\sup_n \left[\frac{|\lambda_n|^2}{|\operatorname{Im}\lambda_n|}\left(\|\psi_n\| - \|F(\lambda_n)\psi_n\|\right)\right] < \infty. \tag{2.14}$$

Then $\psi \in \mathfrak{N}_{\lambda_0,\,\infty}$, *the sequence* $\{F(\lambda_n)\psi_n\}$ *converges weakly to* $V_\infty\psi$ *and*

$$\lim_{n\to\infty}\left[\frac{|\lambda_n|^2}{|\operatorname{Im}\lambda_n|}\left(\|\psi_n\|^2 - \|F(\lambda_n)\psi_n\|\right)^2\right] \geqslant \frac{1}{|\operatorname{Im}\lambda_0|}\|\hat{P}\widetilde{A}(V_\infty\psi - \psi)\|^2. \tag{2.15}$$

If the sequence $\{\psi_n\}$ *converges strongly to* ψ, *then*

$$\lim_{n\to\infty} F(\lambda_n)\psi_n = \lim_{n\to\infty} F(\lambda_n)\psi = V_\infty\psi.$$

Proof. Consider again the sequence $\{\widetilde{A}\widetilde{g}_n\}$ where $\widetilde{g}_n \in \widetilde{\mathfrak{L}}_{\lambda_n}$ and $P\widetilde{g}_n = F(\lambda_n)\psi_n - \psi_n$ $(n = 1, 2, \cdots)$. Because

$$P\widetilde{A}\widetilde{g}_n = \lambda_0 F(\lambda_n)\psi_n - \overline{\lambda}_0\psi_n \quad (n = 1, 2, \ldots), \tag{2.16}$$

$\{P\widetilde{A}\widetilde{g}_n\}$ is bounded. According to (1.5)

$$\|\hat{P}\widetilde{A}\widetilde{g}_n\|^2 = \operatorname{Im}\lambda_0 \frac{|\lambda_n|^2}{\operatorname{Im}\lambda_n}(\|\psi_n\|^2 - \|F(\lambda_n)\psi_n\|^2), \tag{2.17}$$

and therefore by (2.14) $\{\hat{P}\widetilde{A}\widetilde{g}_n\}$ is bounded. Hence $\{\widetilde{A}\widetilde{g}_n\}$ is also bounded.

We now observe that according to (1.2)

$$\hat{P}\widetilde{g}_n = \frac{1}{\lambda_n}\hat{P}\widetilde{A}\widetilde{g}_n \quad (n = 1, 2, \ldots),$$

so that

$$\lim_{n\to\infty} \hat{P}\widetilde{g}_n = 0. \tag{2.18}$$

Keeping in mind the property of weak compactness of bounded sets in a Hilbert space, we now investigate an arbitrary weakly convergent subsequence $\{\widetilde{A}\widetilde{g}_{n_k}\}$ of the bounded sequence $\{\widetilde{A}\widetilde{g}_n\}$. Then as $k \to \infty$

$$\widetilde{A}\widetilde{g}_{n_k} \rightharpoonup \widetilde{h}, \tag{2.19}$$

where $\widetilde{h}$ is some element from $\widetilde{H}$. Clearly

$$P\widetilde{A}\widetilde{g}_{n_k} \rightharpoonup P\widetilde{h}.$$

Taking (2.16) into account, we conclude that as $k \to \infty$

$$F(\lambda_{n_k})\psi_{n_k} \rightharpoonup \varphi, \tag{2.20}$$

where

$$\varphi = \frac{1}{\lambda_0}(P\widetilde{h} + \bar{\lambda}_0\psi).$$

It is clear that $\phi \in \mathfrak{N}_{\bar{\lambda}_0}$. Since

$$\widetilde{g}_n = F(\lambda_n)\psi_n - \psi_n + \hat{P}\widetilde{g}_n,$$

it follows from (2.18) and (2.20) that

$$\widetilde{g}_{n_k} \rightharpoonup \varphi - \psi \tag{2.21}$$

as $k \to \infty$.

From (2.21) and (2.19) it follows that $\phi - \psi \in D_{\widetilde{A}}$ and $\widetilde{h} = \widetilde{A}(\phi - \psi)$; since also $\phi - \psi \in H$ it follows that $\phi - \psi \in \mathfrak{L}_\infty$. This means that $\psi \in \mathfrak{N}_{\lambda_0,\infty}$, $\phi = V_\infty \psi$ and

$$\widetilde{h} = \widetilde{A}(V_\infty\psi - \psi). \tag{2.22}$$

Moreover,

$$P\widetilde{A}(V_\infty\psi - \psi) = \lambda_0 V_\infty\psi - \bar{\lambda}_0\psi. \tag{2.23}$$

According to (2.19) and (2.22),

$$\widetilde{A}\widetilde{g}_{n_k} \rightharpoonup \widetilde{A}(V_\infty\psi - \psi)$$

as $k \to \infty$. So any weakly converging subsequence of the bounded sequence $\{\widetilde{A}\widetilde{g}_n\}$ has one and the same weak limit . Using Lemma 1.1, we conclude that as $n \to \infty$

$$\widetilde{A}\widetilde{g}_n \rightharpoonup \widetilde{A}(V_\infty\psi - \psi).$$

Therefore

$$P\tilde{A}\tilde{g}_n \rightharpoonup P\tilde{A}(V_\infty\psi - \psi) \tag{2.24}$$

and

$$\hat{P}\tilde{A}\tilde{g}_n \rightharpoonup \hat{P}\tilde{A}(V_\infty\psi - \psi). \tag{2.25}$$

By (2.16) and (2.23) relation (2.24) means that

$$\lambda_0 F(\lambda_n)\psi_n - \bar{\lambda}_0\psi_n \rightharpoonup \lambda_0 V_\infty\psi - \bar{\lambda}_0\psi,$$

and because $\psi_n \rightharpoonup \psi$, it follows that

$$F(\lambda_n)\psi_n \rightharpoonup V_\infty\psi$$

as $n \to \infty$.

Taking into account (2.25) and the well-known estimate of the norm of a weak limit, we get

$$\varliminf_{n\to\infty} \|\hat{P}\tilde{A}\tilde{g}_n\| \geqslant \|\hat{P}\tilde{A}(V_\infty\psi - \psi)\|.$$

Hence by (2.17) we get the inequality (2.15).

If $\{\psi_n\}$ converges strongly to ψ, then

$$\varlimsup_{n\to\infty} \|F(\lambda_n)\psi_n\| \leqslant \lim_{n\to\infty} \|\psi_n\| = \|\psi\| = \|V_\infty\psi\|,$$

and therefore $\{F(\lambda_n)\psi_n\}$, which converges weakly to the element $V_\infty\psi$, also converges strongly to it. It is clear that in this case $\{F(\lambda_n)\psi\}$ also converges strongly to the same limit.

The theorem is thus fully proved.

Observation 1. If in addition we assume that $\lambda_n \in \Pi_\epsilon$ $(n = 1, 2, \cdots)$, * then (2.14) is equivalent to the condition

$$\sup_n [|\lambda_n| (\|\psi_n\| - \|F(\lambda_n)\psi_n\|)] < \infty.$$

Observation 2. If for some sequence of numbers $\lambda_n \in \Pi$ converging to ∞ and for some sequence of elements $\psi_n \in \mathfrak{N}_{\lambda_0}$ converging weakly to ψ

$$\lim_{n\to\infty} \left[\frac{|\lambda_n|^2}{\operatorname{Im}\lambda_n} (\|\psi_n\| - \|F(\lambda_n)\psi_n\|)\right] = 0, \tag{2.26}$$

then

$$\tilde{A}(V_\infty\psi - \psi) = \lambda_0 V_\infty\psi - \bar{\lambda}_0\psi \tag{2.27}$$

and for any $\lambda \in \Pi$

$$F(\lambda)\psi = V_\infty\psi. \tag{2.28}$$

---

* Recall that $\Pi_\epsilon = \{\lambda : \lambda \in \Pi, \epsilon < |\arg\lambda| < \pi - \epsilon\}$, where $0 < \epsilon < \pi/2$.

In fact by (2.26) and (2.15)

$$\hat{P}\tilde{A}(V_\infty\psi - \psi) = 0;$$

that is,

$$\tilde{A}(V_\infty\psi - \psi) \in H.$$

Thus we get (2.27) and (2.28) (cf. [3], §1.3).

Theorem 2.3. *Let the elements* $\phi \in \mathfrak{N}_{\bar{\lambda}_0}$ *and* $\psi \in \mathfrak{N}_{\lambda_0}$ *satisfy the following conditions*: 1) $\|\phi\| = \|\psi\|$; 2) *for some sequence of numbers* $\lambda_n \in \Pi$ *converging to* ∞,

$$\sup_n \left| \frac{\lambda_n^2}{\operatorname{Im}\lambda_n} (F(\lambda_n)\psi - \varphi, \varphi) \right| < \infty. \tag{2.29}$$

Then $\phi - \psi \in \mathfrak{L}_\infty$ *and*

$$\lim_{n\to\infty} F(\lambda_n)\psi = \varphi. \tag{2.30}$$

We omit the details of the proof since it is completely analogous to the the proof of Theorem 1.2. From conditions 1) and 2) it follows that equation (2.30) is valid and

$$\sup_n \left[\frac{|\lambda_n|^2}{|\operatorname{Im}\lambda_n|} (\|\psi\| - \|F(\lambda_n)\psi\|) \right] < \infty.$$

It remains to apply Theorem 2.2.

Observation 1. If $\lambda_n \in \Pi_\epsilon$ $(n = 1, 2, \cdots)$, then (2.29) is equivalent to the condition

$$\sup_n |\lambda_n (F(\lambda_n)\psi - \varphi, \varphi)| < \infty.$$

Observation 2. Condition 2) may be replaced by the following weaker condition: for some sequence of numbers $\lambda_n \in \Pi$ converging to ∞

$$\sup_n \left[\frac{|\lambda_n|^2}{|\operatorname{Im}\lambda_n|} \operatorname{Re}(\varphi - F(\lambda_n)\psi, \varphi) \right] < \infty.$$

2. We now introduce the resolvent of the operator $\tilde{A}$,

$$\tilde{R}_\lambda = (\tilde{A} - \lambda E)^{-1} \quad (\operatorname{Im}\lambda \neq 0),$$

and also the generalized resolvent, generated by $\tilde{A}$, of the operator A,

$$R_\lambda = P\tilde{R}_\lambda|_H \quad (\operatorname{Im}\lambda \neq 0). \tag{2.31}$$

Recall now some of the properties of the resolvent and the generalized resolvent.

For any $\tilde{h} \in \tilde{H}$

$$\lim_{\substack{\lambda\to\infty \\ \lambda\in\Pi_\varepsilon}} (-\lambda\tilde{R}_\lambda\tilde{h}) = \tilde{h}. \tag{2.32}$$

If $\tilde{h} \in D_{\tilde{A}}$, then

$$\tilde{A}\tilde{R}_\lambda\tilde{h} = \tilde{R}_\lambda\tilde{A}\tilde{h},$$

so that if we replace $\tilde{h}$ by $\widetilde{\tilde{A}\tilde{h}}$ in (2.32) we get

$$\lim_{\substack{\lambda\to\infty\\ \lambda\in\Pi_\varepsilon}} (-\lambda\tilde{A}\tilde{R}_\lambda\tilde{h}) = \tilde{A}\tilde{h}.$$

From (2.32) and (2.31) it follows that for any $h \in H$

$$\lim_{\substack{\lambda\to\infty\\ \lambda\in\Pi_\varepsilon}} (-\lambda R_\lambda h) = h.$$

Moreover, for any nonreal λ

$$\tilde{\mathfrak{L}}_\lambda = \tilde{R}_\lambda H, \quad \mathfrak{L}_\lambda = R_\lambda H, \tag{2.33}$$

and equation (0.1) can be rewritten in the form

$$(B_\lambda - \lambda E)^{-1} = R_\lambda. \tag{2.34}$$

Theorem 2.4. *Let ψ be an arbitrary element from $\mathfrak{N}_{\lambda_0,\infty}$, and let the vector-function $\psi(\lambda)$ $(\lambda \in \Pi)$ with values in $\mathfrak{N}_{\lambda_0}$ be given by formula*

$$\psi(\lambda) = \frac{\lambda}{\lambda_0 - \bar{\lambda}_0} P_{\mathfrak{N}_{\lambda_0}} (A^* - \lambda_0 E) R_\lambda (\psi - V_\infty\psi). \tag{2.35}$$

Then

$$\lim_{\substack{\lambda\to\infty\\ \lambda\in\Pi_\varepsilon}} \psi(\lambda) = \psi, \tag{2.36}$$

$$\lim_{\substack{\lambda\to\infty\\ \lambda\in\Pi_\varepsilon}} F(\lambda)\psi(\lambda) = V_\infty\psi \tag{2.37}$$

and the limit

$$\lim_{\substack{\lambda,\,\zeta\to\infty\\ \lambda,\,\zeta\in\Pi_\varepsilon}} \left\{\frac{\lambda\bar{\zeta}}{\lambda - \bar{\zeta}} [(\psi(\lambda), \psi(\zeta)) - (F(\lambda)\psi(\lambda), F(\zeta)\psi(\zeta))]\right\} \tag{2.38}$$

exists and is finite.

Proof. We assume

$$g = V_\infty\psi - \psi, \tag{2.39}$$

$$\tilde{g}(\lambda) = -\lambda\tilde{R}_\lambda g, \tag{2.40}$$

$$g(\lambda) = P\tilde{g}(\lambda) = -\lambda R_\lambda g \quad (\lambda \in \Pi). \tag{2.41}$$

By comparing (2.35), (2.39) and (2.41) we see that

$$\psi(\lambda) = \frac{1}{\lambda_0 - \bar{\lambda}_0} P_{\mathfrak{N}_{\lambda_0}} (A^* - \lambda_0 E) g(\lambda). \tag{2.42}$$

According to the properties of the resolvent and the generalized resolvent mentioned above it follows from (2.40) and (2.41) that

$$\lim_{\substack{\lambda\to\infty\\ \lambda\in\Pi_\varepsilon}} \tilde{g}(\lambda) = g, \quad \lim_{\substack{\lambda\to\infty\\ \lambda\in\Pi_\varepsilon}} \tilde{A}\tilde{g}(\lambda) = \tilde{A}g, \tag{2.43}$$

$$\lim_{\substack{\lambda\to\infty \\ \lambda\in\Pi_\varepsilon}} g(\lambda) = g. \tag{2.44}$$

Also, for any $\tilde{g} \in D_{\tilde{A}}$

$$P\tilde{g} \in D_{A^*}, \quad P\tilde{A}\tilde{g} = A^*P\tilde{g}.$$

Hence by (2.43)

$$\lim_{\substack{\lambda\to\infty \\ \lambda\in\Pi_\varepsilon}} A^*g(\lambda) = A^*g. \tag{2.45}$$

Taking (2.42), (2.44) and (2.45) into account, we get

$$\lim_{\substack{\lambda\to\infty \\ \lambda\in\Pi_\varepsilon}} \psi(\lambda) = \frac{1}{\lambda_0 - \bar{\lambda}_0} P_{\mathfrak{N}_{\lambda_0}} (A^* - \lambda_0 E)\, g,$$

and since according to (2.39)

$$(A^* - \lambda_0 E)\, g = (\lambda_0 - \bar{\lambda}_0)\, \psi,$$

we have proved equation (2.36).

From (2.33) and (2.41) it follows that $g(\lambda) \in \mathfrak{L}_\lambda$ for any $\lambda \in \Pi$, and therefore, according to formula (0.2), $g(\lambda)$ has the representation

$$g(\lambda) = f(\lambda) + F(\lambda)\psi_1(\lambda) - \psi_1(\lambda) \quad (\lambda \in \Pi), \tag{2.46}$$

where $f(\lambda) \in D_A$ and $\psi_1(\lambda) \in \mathfrak{N}_{\lambda_0}$. In the expansion (2.46) the component $\psi_1(\lambda)$ can be expressed in terms of $g(\lambda)$ by the formula

$$\psi_1(\lambda) = \frac{1}{\lambda_0 - \bar{\lambda}_0} P_{\mathfrak{N}_{\lambda_0}} (A^* - \lambda_0 E)\, g(\lambda),$$

so that by (2.42)

$$\psi_1(\lambda) = \psi(\lambda)$$

and therefore we can rewrite (2.46) in the form

$$g(\lambda) = f(\lambda) + F(\lambda)\psi(\lambda) - \psi(\lambda) \quad (\lambda \in \Pi). \tag{2.47}$$

Hence it follows in particular that

$$F(\lambda)\,\psi(\lambda) = \frac{1}{\lambda_0 - \bar{\lambda}_0} P_{\mathfrak{N}_{\bar{\lambda}_0}} (A^* - \bar{\lambda}_0 E)\, g(\lambda) \quad (\lambda \in \Pi).$$

Taking (2.44), (2.45) and (2.39) into account, we see that (2.37) is correct.*

According to (1.5), for any $\lambda, \zeta \in \Pi$ we have the equation

$$\frac{\lambda\bar{\zeta}}{\lambda - \bar{\zeta}}[(\psi(\lambda), \psi(\zeta)) - (F(\lambda)\psi(\lambda), F(\zeta)\psi(\zeta))] = \frac{1}{\lambda_0 - \bar{\lambda}_0}(\hat{P}\widetilde{A}\tilde{g}(\lambda), \hat{P}\widetilde{A}\tilde{g}(\zeta)).$$

From this equation and (2.43) we get the existence of the finite limit (2.38). Thus the theorem is proved.

* On the other hand, it would be possible first to prove the existence of (2.38) and then show that (2.37) is valid using Theorem 2.1.

Corollary 1. *If* $\psi \in \mathfrak{N}_{\lambda_0, \infty}$, *then*

$$\lim_{\substack{\lambda\to\infty \\ \lambda\in\Pi_\varepsilon}} F(\lambda)\psi = V_\infty\psi. \tag{2.48}$$

In fact (2.48) follows immediately from (2.36) and (2.37), since $\|F(\lambda)\| \leq 1$ for any $\lambda \in \Pi$.

Corollary 2. *Under the conditions of Theorem* 2.4 *the limit*

$$\lim_{\substack{\lambda\to\infty \\ \lambda\in\Pi_\varepsilon}} \left[\frac{|\lambda|^2}{\operatorname{Im}\lambda}(\|\psi(\lambda)\| - \|F(\lambda)\psi(\lambda)\|)\right] \tag{2.49}$$

exists and is finite.

In fact, since the finite limit (2.38) exists, the limit

$$\lim_{\substack{\lambda\to\infty \\ \lambda\in\Pi_\varepsilon}} \left[\frac{|\lambda|^2}{\lambda-\bar{\lambda}}(\|\psi(\lambda)\|^2 - \|F(\lambda)\psi(\lambda)\|^2)\right] \tag{2.50}$$

exists and coincides with it. Moreover,

$$\lim_{\substack{\lambda\to\infty \\ \lambda\in\Pi_\varepsilon}} (\|\psi(\lambda)\| + \|F(\lambda)\psi(\lambda)\|) = 2\|\psi\|,$$

so that for $\psi \neq 0$ the limit (2.49) exists along with (2.50); if $\psi = 0$, then $\psi(\lambda) = 0$ and (2.49) exists trivially.

3. We will now assume that the operator A in H is the orthogonal sum of two maximal symmetric operators A_1 and A_2 with regions of regularity which do not coincide.* A_1 and A_2 operate in the Hilbert spaces H_1 and H_2 respectively; moreover $H = H_1 \oplus H_2$. For definiteness let the half-plane Π serve as the region of regularity for A_2. Then for any $\lambda \in \Pi$

$$\mathfrak{N}_\lambda = H_1 \ominus (A_1 - \lambda E)D_{A_1}$$

and

$$\mathfrak{N}_{\bar{\lambda}} = H_2 \ominus (A_2 - \bar{\lambda}E)D_{A_2}.$$

We will show that in the case under consideration formula (2.35) reduces to a much simpler form.

Let $\psi \in \mathfrak{N}_{\lambda_0}$ and $\lambda \in \Pi$. According to (2.34) the element $R_\lambda\psi$ is uniquely defined by the conditions

$$R_\lambda\psi \in \mathfrak{L}_\lambda \tag{2.51}$$

and

$$(A^* - \lambda E)R_\lambda\psi = \psi. \tag{2.52}$$

---

* The region of regularity of a maximal symmetric operator is one of the half-planes (upper or lower).

By (2.51) and (0.2)

$$R_\lambda\psi = f + F(\lambda)\psi_1 - \psi_1,$$

where $f \in D_A$ and $\psi_1 \in \mathfrak{N}_{\lambda_0}$. It is necessary to define the elements f and ψ_1 in such a way that condition (2.52) in the following form is fulfilled:

$$(A - \lambda E)f + (\lambda_0 - \lambda)F(\lambda)\psi_1 - (\overline{\lambda}_0 - \lambda)\psi_1 = \psi.$$

It is easy to verify that this equation holds if we assume

$$\psi_1 = \frac{1}{\lambda - \overline{\lambda}_0}\psi$$

and

$$f = (\lambda - \lambda_0)(A_2 - \lambda E)^{-1} F(\lambda)\psi_1 = \frac{\lambda - \lambda_0}{\lambda - \overline{\lambda}_0}(A_2 - \lambda E)^{-1} F(\lambda)\psi;$$

Here we consider that $F(\lambda)\psi_1 \in H_2$ and $(A_2 - \lambda E)D_{A_2} = H_2$. So

$$R_\lambda\psi = \frac{\lambda - \lambda_0}{\lambda - \overline{\lambda}_0}(A_2 - \lambda E)^{-1} F(\lambda)\psi + \frac{1}{\lambda - \overline{\lambda}_0}F(\lambda)\psi - \frac{1}{\lambda - \overline{\lambda}_0}\psi.$$

Hence

$$P_{\mathfrak{N}_{\lambda_0}}(A^* - \lambda_0 E)R_\lambda\psi = \frac{\lambda_0 - \overline{\lambda}_0}{\lambda - \overline{\lambda}_0}\psi. \tag{2.53}$$

For any $\phi \in \mathfrak{N}_{\overline{\lambda}_0}$ and any $\lambda \in \Pi$

$$R_\lambda\varphi = (A_2 - \lambda E)^{-1}\varphi$$

and consequently

$$P_{\mathfrak{N}_{\lambda_0}}(A^* - \lambda_0 E)R_\lambda\varphi = 0. \tag{2.54}$$

Now let $\psi \in \mathfrak{N}_{\lambda_0, \infty}$. By (2.53) and (2.54) the vector-function $\psi(\lambda)$ ($\lambda \in \Pi$), given by (2.35), takes the following form:

$$\psi(\lambda) = \frac{\lambda}{\lambda - \overline{\lambda}_0}\psi.$$

Using Theorem 2.4 we get the following result.

Lemma 2.1. *If the operator A is the orthogonal sum of two maximal symmetric operators, then for any $\psi \in \mathfrak{N}_{\lambda_0, \infty}$ the limit*

$$\lim_{\substack{\lambda, \zeta \to \infty \\ \lambda, \zeta \in \Pi_\varepsilon}} \left\{\frac{\lambda\overline{\zeta}}{\lambda - \overline{\zeta}}\,[(\psi, \psi) - (F(\lambda)\psi, F(\zeta)\psi)]\right\}^* \tag{2.55}$$

exists and is finite.

Further on in §4 we will show that (2.55) exists also in the case of an arbitrary symmetric operator A for any element $\psi \in \mathfrak{N}_{\lambda_0, \infty}$.

---

* We assumed previously that the maximal symmetric operators A_1 and A_2 have different regions of regularity. However, the lemma is still true even without this assumption. In fact, if the same half-plane serves as the region of regularity of both operators, then the operator $A = A_1 \oplus A_2$ is also maximal and in this case $\mathfrak{N}_{\lambda_0, \infty}$ consists only of the zero vector.

§3. On boundary properties of analytic operator functions

Let $\Phi(\lambda)$ $(\lambda \in \Pi)$ be an analytic operator function in the half-plane Π the values of which are nonexpansive linear operators from the Hilbert space $\mathfrak{N}$ into the Hilbert space $\mathfrak{N}'$. We will prove some propositions on the behavior of the operator function $\Phi(\lambda)$ as $\lambda \to \infty$, and then we will rephrase them for the case when $\lambda \to \mu$ where μ is an arbitrary real number. These propositions are closely related to results set forth in the previous part of this article and will be used in §4 to describe the linear manifolds $\mathfrak{L}_\mu$ and $\mathfrak{L}_\infty$ in terms of the operator function $F(\lambda)$.

Theorem 3.1. *For some sequence of numbers $\lambda_n \in \Pi$ converging to ∞ and for some sequence of elements $f_n \in \mathfrak{N}$ converging weakly to the element f let*

$$\sup_n \left[\frac{|\lambda_n|^2}{|\operatorname{Im}\lambda_n|}(\|f_n\| - \|\Phi(\lambda_n) f_n\|)\right] < \infty.$$

Then the limit

$$\lim_{\substack{\lambda,\zeta\to\infty \\ \lambda,\zeta\in\Pi_\varepsilon}} \left\{\frac{\lambda\bar{\zeta}}{\lambda-\bar{\zeta}}[(f, f) - (\Phi(\lambda) f, \Phi(\zeta) f)]\right\} \tag{3.1}$$

is finite.

Proof. We introduce a symmetric operator A in H like the one of §2.3, and we assume in addition that

$$\dim \mathfrak{N}_{\lambda_0} = \dim \mathfrak{N}, \quad \dim \mathfrak{N}_{\bar{\lambda}_0} = \dim \mathfrak{N}'.$$

Let U and W be isometric operators mapping the defect subspaces $\mathfrak{N}_{\lambda_0}$ and $\mathfrak{N}_{\bar{\lambda}_0}$ of the operator A onto $\mathfrak{N}$ and $\mathfrak{N}'$ respectively.

We will assume that

$$F(\lambda) = W^{-1}\Phi(\lambda) U \quad (\lambda \in \Pi).$$

The operator function $F(\lambda)$ so defined is analytic in the half-plane Π, and $\|F(\lambda)\| \le 1$ for any $\lambda \in \Pi$. Therefore according to the converse proposition associated with (0.2) the operator function $F(\lambda)$ corresponds to some selfadjoint extension $\widetilde{A}$ of A operating in the Hilbert space $\widetilde{H} \supset H$.

We will assume $\psi_n = Uf_n$ $(n = 1, 2, \cdots)$ and $\psi = Uf$. From the conditions of the theorem it follows that Theorem 2.2 is applicable to $\{\lambda_n\}$, $\{\psi_n\}$ and the element ψ, and therefore $\psi \in \mathfrak{N}_{\lambda_0, \infty}$. Then by Lemma 2.1 the finite limit (2.55) exists, and this means the finite limit (3.1) exists. Thus the theorem is proved.

Corollary 1. *Under the conditions of Theorem 3.1 the strong limit*

$$\lim_{\substack{\lambda\to\infty \\ \lambda\in\Pi_\varepsilon}} \Phi(\lambda) f = g, \tag{3.2}$$

exists; also

$$\|g\| = \|f\|. \tag{3.3}$$

In fact, by virtue of the theorem

$$\lim_{\substack{\lambda,\,\zeta\to\infty \\ \lambda,\,\zeta\in\Pi_\varepsilon}} (\Phi(\lambda) f, \Phi(\zeta) f) = (f, f).$$

Hence we have the existence of the limit (3.2) and equation (3.3).

Corollary 2. *Under the conditions of Theorem* 3.1 *the finite limit*

$$\lim_{\substack{\lambda\to\infty \\ \lambda\in\Pi_\varepsilon}} \left[\frac{|\lambda|^2}{\operatorname{Im}\lambda}(\|f\| - \|\Phi(\lambda) f\|)\right] \tag{3.4}$$

exists.

In fact, from Theorem 3.1 it follows that the limit

$$\lim_{\substack{\lambda\to\infty \\ \lambda\in\Pi_\varepsilon}} \left[\frac{|\lambda^2|}{\operatorname{Im}\lambda}(\|f\|^2 - \|\Phi(\lambda) f\|^2)\right],$$

exists and if finite, which is equivalent to the existence of the finite limit (3.4).

Corollary 3. *Under the conditions of Theorem* 3.1 *the limit*

$$\lim_{\substack{\lambda\to\infty \\ \lambda\in\Pi_\varepsilon}} [\lambda(\Phi(\lambda) f - g, g)], \tag{3.5}$$

exists and is finite; g *is defined by* (3.2). *The limit* (3.5) *coincides with the limit* (3.1).

In fact, for any fixed $\lambda\in\Pi$, according to Corollary 1, the limit

$$\lim_{\substack{\zeta\to\infty \\ \zeta\in\Pi_\varepsilon}} \left\{\frac{\lambda\bar{\zeta}}{\lambda-\bar{\zeta}}[(f, f) - (\Phi(\lambda) f, \Phi(\zeta) g)]\right\}$$
$$= -\lambda[(f, f) - (\Phi(\lambda) f, g)] = \lambda(\Phi(\lambda) f - g, g)$$

exists and is finite.

On the basis of a well-known theorem on the equality of double and iterated limits we conclude that limit (3.5) exists along with (3.1) and that both are equal.

Observation 1. If for some sequence of numbers $\lambda_n\in\Pi$ converging to ∞ and for some sequence of elements $f_n\in\mathfrak{N}$ converging weakly to the element f,

$$\lim_{n\to\infty}\left[\frac{|\lambda_n|^2}{\operatorname{Im}\lambda_n}(\|f_n\| - \|\Phi(\lambda_n) f_n\|)\right] = 0,$$

then $\Phi(\lambda) f (\lambda \in \Pi)$ does not depend on λ and $\|\Phi(\lambda)f\| = \|f\|$.*

This follows from the proof of the theorem and from Observation 2 to Theorem 2.2.

Observation 2. If a is the common value of (3.1) and (3.5) and b is the limit (3.4), then

$$a = -i\|f\|b, \quad \operatorname{Re} a = 0, \quad \operatorname{Im} a \cdot \operatorname{Im} \lambda_0 \leqslant 0.$$

According to Observation 1, $a = 0$ if and only if $\Phi(\lambda)f$ $(\lambda \in \Pi)$ does not depend on λ and $\|\Phi(\lambda)f\| = \|f\|$.

By applying Corollary 3 and Observation 2 to the frequent case when $\dim \mathfrak{N} = \dim \mathfrak{N}' = 1$, we get the following proposition.

Let the function $\omega(\lambda)$ $(\lambda \in \Pi)$ *be analytic in the half-plane* Π *and let* $|\omega(\lambda)| < 1$. *If for some sequence of numbers* $\lambda_n \in \Pi$ *converging to* ∞, $\lim_{n\to\infty} \omega(\lambda_n) = 1$, *and*

$$\sup_n \left[\frac{|\lambda_n|^2}{|\operatorname{Im} \lambda_n|}(1 - |\omega(\lambda_n)|)\right] < \infty,$$

then the limit

$$\lim_{\substack{\lambda\to\infty \\ \lambda \in \Pi_\varepsilon}} [\lambda(\omega(\lambda) - 1)] = l$$

exists and is finite; moreover $\operatorname{Re} l = 0$ *and* $\operatorname{Im} l \cdot \operatorname{Im} \lambda_0 < 0$.

If we reformulate this proposition for functions analytic in the unit circle, it coincides with a theorem of Carathéodory ([5], Russian pp. 70, 114–115).

We now supplement Corollaries 1 and 3 with the following proposition.

Theorem 3.2. *Under the conditions of Theorem* 3.1 *for an element* g *defined by* (3.2) *we have the equation*

$$\lim_{\substack{\lambda\to\infty \\ \lambda \in \Pi_\varepsilon}} (|\lambda|^{1/2} \|\Phi(\lambda) f - g\|) = 0^{**}.$$

Proof. Consider the equation

$$\|\Phi(\lambda) f - g\|^2 = [\|\Phi(\lambda) f\|^2 - \|g\|^2] + [\|g\|^2 - (\Phi(\lambda) f, g)] + (\|g\|^2 - (g, \Phi(\lambda) f)]. \tag{3.6}$$

Taking the equation $\|g\| = \|f\|$ into account and as before letting a be the pure imaginary number which is equal to both the limits (3.1) and (3.5), we have

---

* If for even one $\lambda \in \Pi$ $\|\Phi(\lambda)f\| = \|f\|$, then it follows by the maximum modulus principle that $\Phi(\lambda)f$ $(\lambda \in \Pi)$ does not depend on λ.

**In the frequent case that $\dim \mathfrak{N} = \dim \mathfrak{N}' = 1$, Theorem 3.2 follows immediately from Corollary 3 and also from the cited result of Carathéodory.

$$\left.\begin{aligned} \|\Phi(\lambda) f\|^2 - \|g\|^2 &= \frac{\lambda - \bar{\lambda}}{|\lambda|^2}(-a + \alpha(\lambda)), \\ \|g\|^2 - (\Phi(\lambda) f, g) &= \frac{1}{\lambda}(-a + \beta(\lambda)), \\ \|g\|^2 - (g, \Phi(\lambda) f) &= \frac{1}{\bar{\lambda}}(a + \overline{\beta(\lambda)}), \end{aligned}\right\} \tag{3.7}$$

where

$$\lim_{\substack{\lambda\to\infty \\ \lambda\in\Pi_\varepsilon}} \alpha(\lambda) = \lim_{\substack{\lambda\to\infty \\ \lambda\in\Pi_\varepsilon}} \beta(\lambda) = 0.$$

By (3.6) and (3.7)

$$\|\Phi(\lambda) f - g\|^2 = \frac{\lambda - \bar{\lambda}}{|\lambda|^2}\alpha(\lambda) + \frac{1}{\lambda}\beta(\lambda) + \frac{1}{\bar{\lambda}}\overline{\beta(\lambda)}$$

and consequently

$$\lim_{\substack{\lambda\to\infty \\ \lambda\in\Pi_\varepsilon}} (|\lambda| \|\Phi(\lambda) f - g\|^2) = 0.$$

Thus the theorem is proved.

The propositions established here can be reformulated so as to be applicable to the case when $\lambda \to \mu$, where μ is an arbitrary real number. It is sufficient to assume

$$\lambda' = \frac{1}{\mu - \lambda},$$

$$\Phi_1(\lambda') = \Phi(\lambda) \qquad (\lambda \in \Pi)$$

and use the results related to the behavior of the operator function $\Phi_1(\lambda')$ $(\lambda' \in \Pi)$ as $\lambda' \to \infty$.

Instead of Theorem 3.1 we get the following proposition.

Theorem 3.1′. *For some sequence of numbers $\lambda_n \in \Pi$ converging to μ and for some sequence of elements $f_n \in \mathfrak{N}$ converging weakly to the element f, let*

$$\sup_n \frac{\|f_n\| - \|\Phi(\lambda_n) f_n\|}{|\operatorname{Im} \lambda_n|} < \infty.$$

Then the limit

$$\lim_{\substack{\lambda,\zeta\to\mu \\ \lambda,\zeta\in\Pi_{\mu,\varepsilon}}} \frac{(f, f) - (\Phi(\lambda) f, \Phi(\zeta) f)}{\lambda - \bar{\zeta}} \tag{3.8}$$

exists and is finite.

We combine in the form of a corollary to Theorem 3.1′ the propositions analogous to Corollaries 1–3 of Theorem 3.1.

Corollary. *Under the conditions of Theorem 3.1′ the strong limit*

$$\lim_{\substack{\lambda\to\mu \\ \lambda\in\Pi_{\mu,\varepsilon}}} \Phi(\lambda) f = g \tag{3.9}$$

and the finite limits

$$\lim_{\substack{\lambda\to\mu \\ \lambda\in\Pi_{\mu,\varepsilon}}} \frac{\|f\| - \|\Phi(\lambda) f\|}{\operatorname{Im}\lambda},$$

$$\lim_{\substack{\lambda\to\mu \\ \lambda\in\Pi_{\mu,\varepsilon}}} \frac{(g - \Phi(\lambda) f, g)}{\lambda-\mu} \tag{3.10}$$

exist; here $\|g\| = \|f\|$ *and* (3.10) *coincides with* (3.8).

The following proposition is a reformulation of Theorem 3.2.

Theorem 3.2$'$. *Under the conditions of Theorem* 3.1$'$, *for an element* g *defined by* (3.9) *we have the equation*

$$\lim_{\substack{\lambda\to\mu \\ \lambda\in\Pi_{\mu,\varepsilon}}} \frac{\|\Phi(\lambda) f - g\|}{|\lambda-\mu|^{\frac{1}{2}}} = 0.$$

§4. Description of the linear manifolds $\mathfrak{L}_\mu$ and $\mathfrak{L}_\infty$

1. We turn to the study of the linear manifold $\mathfrak{L}_\mu$ for any real number μ. The following theorem strengthens and at the same time simplifies proposition II of the Introduction.

Theorem 4.1. *Let* $\psi \in \mathfrak{N}_{\lambda_0,\mu}$. *Then the limit*

$$\lim_{\substack{\lambda,\zeta\to\mu \\ \lambda,\zeta\in\Pi_{\mu,\varepsilon}}} \frac{(\psi,\psi) - (F(\lambda)\psi, F(\zeta)\psi)}{\lambda-\bar{\zeta}} \tag{4.1}$$

exists and is finite.

Proof. According to proposition II of the Introduction, for any sequence of numbers $\lambda_n \in \Pi_{\mu,\epsilon}$ converging to μ there exists a sequence of elements $\psi_n \in \mathfrak{N}_{\lambda_0}$ strongly converging to ψ such that the double sequence of numbers

$$\omega_{mn} = \frac{1}{\lambda_n - \bar{\lambda}_m} [(\psi_n, \psi_m) - (F(\lambda_n)\psi_n, F(\lambda_m)\psi_m)]$$

has a finite limit a as $m, n \to \infty$. It is clear that

$$\lim_{n\to\infty} \omega_{nn} = a. \tag{4.2}$$

We can assume $\psi \neq 0$, because for $\psi = 0$ the theorem is trivial. But if $\psi \neq 0$, then by (4.2) we have the equation

$$\lim_{n\to\infty} \frac{\|\psi_n\| - \|F(\lambda_n)\psi_n\|}{\operatorname{Im}\lambda_n} = \frac{ai}{\|\psi\|}$$

By applying Theorem 3.1$'$ we see that (4.1) exists and is finite.

Corollary. *If* $\psi \in \mathfrak{N}_{\lambda_0,\mu}$, *then the limits*

$$\lim_{\substack{\lambda \to \mu \\ \lambda \in \Pi_{\mu,\varepsilon}}} \frac{\|\psi\| - \|F(\lambda)\psi\|}{\operatorname{Im}\lambda}$$

and

$$\lim_{\substack{\lambda \to \mu \\ \lambda \in \Pi_{\mu,\varepsilon}}} \frac{(F(\lambda)\psi - V_\mu\psi, V_\mu\psi)}{\lambda - \mu}$$

exist and are finite.

This proposition is analogous to the Corollary to Theorem 3.1′. We need only consider that for any $\psi \in \mathfrak{N}_{\lambda_0,\mu}$, as was mentioned in the Introduction,

$$\lim_{\substack{\lambda \to \mu \\ \lambda \in \Pi_{\mu,\varepsilon}}} F(\lambda)\psi = V_\mu\psi.$$

By comparing Theorem 1.1 and the Corollary to Theorem 4.1 we get the following description of the linear manifold $\mathfrak{N}_{\lambda_0,\mu}$ for any real μ.

Theorem 4.2. *In order that the element $\psi \in \mathfrak{N}_{\lambda_0}$ belong to the linear manifold $\mathfrak{N}_{\lambda_0,\mu}$ it is necessary and sufficient that the limit*

$$\lim_{\substack{\lambda \to \mu \\ \lambda \in \Pi_{\mu,\varepsilon}}} \frac{\|\psi\| - \|F(\lambda)\psi\|}{\operatorname{Im}\lambda}$$

exist and be finite.

Formula (0.3), Theorem 1.2 (together with Observation 1), and the Corollary to Theorem 4.1 lead to the following description of the linear manifold $\mathfrak{L}_\mu$.

Theorem 4.3. *The linear manifold $\mathfrak{L}_\mu$ consists of all elements g of the form*

$$g = f + \phi - \psi,$$

where f is an arbitrary element from D_A and the elements $\phi \in \mathfrak{N}_{\overline{\lambda}_0}$ and $\psi \in \mathfrak{N}_{\lambda_0}$ satisfy the conditions: 1) $\|\phi\| = \|\psi\|$; 2) *the limit*

$$\lim_{\substack{\lambda \to \mu \\ \lambda \in \Pi_{\mu,\varepsilon}}} \frac{(F(\lambda)\psi - \varphi, \varphi)}{\lambda - \mu}$$

exists and is finite.

2. Again consider the linear manifold

$$\mathfrak{L}_\infty = D_{\widetilde{A}} \cap H.$$

Taking into account Theorem 2.4, its Corollary 2, and Theorem 3.1, we get the following proposition

Theorem 4.4. *If $\psi \in \mathfrak{N}_{\lambda_0,\infty}$, then the limit*

$$\lim_{\substack{\lambda,\zeta\to\infty\\ \lambda,\zeta\in\Pi_\varepsilon}} \left\{\frac{\lambda\bar{\zeta}}{\lambda-\bar{\zeta}}\,[(\psi,\psi)-(F(\lambda)\psi, F(\zeta)\psi)]\right\}$$

exists and is finite.

Corollary. *If $\psi \in \overline{\mathfrak{N}_{\lambda_0,\infty}}$, then the limits*

$$\lim_{\substack{\lambda\to\infty\\ \lambda\in\Pi_\varepsilon}} \left[\frac{|\lambda|^2}{\operatorname{Im}\lambda}(\|\psi\|-\|F(\lambda)\psi\|)\right]$$

and

$$\lim_{\substack{\lambda\to\infty\\ \lambda\in\Pi_\varepsilon}} [\lambda(F(\lambda)\psi - V_\infty\psi, V_\infty\psi)]$$

exist and are finite.

The situation here is similar to that for Corollaries 2 and 3 of Theorem 3.1. It is also necessary to consider that, according to Corollary 1 of Theorem 2.4, for any $\psi \in \mathfrak{N}_{\lambda_0,\infty}$

$$\lim_{\substack{\lambda\to\infty\\ \lambda\in\Pi_\varepsilon}} F(\lambda)\psi = V_\infty\psi.$$

By Theorem 2.2 and the Corollary to Theorem 4.4 we get a description of the linear manifold $\mathfrak{N}_{\lambda_0,\infty}$ in terms of the operator function $F(\lambda)$.

Theorem 4.5. *In order that $\psi \in \mathfrak{N}_{\lambda_0}$ belong to the linear manifold $\mathfrak{N}_{\lambda_0,\infty}$ it is necessary and sufficient that the limit*

$$\lim_{\substack{\lambda\to\infty\\ \lambda\in\Pi_\varepsilon}} \left[\frac{|\lambda|^2}{\operatorname{Im}\lambda}(\|\psi\|-\|F(\lambda)\psi\|)\right]$$

exist and be finite.

Formula (2.1), Theorem 2.3 along with Observation 1, and the Corollary to Theorem 4.4 allow us to the describe the linear manifold $\mathfrak{L}_\infty$ in the following manner.

Theorem 4.6. *The linear manifold $\mathfrak{L}_\infty$ consists of all elements g of the form*

$$g = f + \phi - \psi,$$

where f is some arbitrary element form D_A and the elements $\phi \in \mathfrak{N}_{\bar{\lambda}_0}$ and $\psi \in \mathfrak{N}_{\lambda_0}$ satisfy the conditions: 1) $||\phi|| = ||\psi||$; 2) *the limit*

$$\lim_{\substack{\lambda\to\infty\\ \lambda\in\Pi_\varepsilon}} [\lambda(F(\lambda)\psi - \varphi, \varphi)]$$

exists and is finite.

3. Recall that according to M. A. Naĭmark [1] the selfadjoint operator $\tilde{A} \supset A$ operating in $\tilde{H} \supset H$ is called an extension of the second kind for A if

$$D_{\widetilde{A}} \cap H = D_A.$$

In our notation this condition takes the form

$$\mathfrak{L}_\infty = D_A.$$

Formula (2.1) shows that $\widetilde{A}$ is a selfadjoint extension of the second kind for A if and only if $\mathfrak{N}_{\lambda_0,\infty}$ consists only of the zero element. We characterize this case with the help of the operator function $F(\lambda)$ ($\lambda \in \Pi$) corresponding to the operator $\widetilde{A}$. So we assume that $F(\lambda)$ ($\lambda \in \Pi$) is an operator function which is analytic in the half-plane Π and whose values are linear nonexpansive operators from $\mathfrak{N}_{\lambda_0}$ into $\mathfrak{N}_{\overline{\lambda}_0}$.

Theorem 4.7. *In order that the operator function $F(\lambda)$ ($\lambda \in \Pi$) correspond to some selfadjoint extension of the second kind of the operator A, it is necessary and sufficient that the following condition be fulfilled: for any nonzero element $\psi \in \mathfrak{N}_{\lambda_0}$*

$$\lim_{\substack{\lambda\to\infty \\ \lambda\in\Pi_\varepsilon}} [|\lambda|(\|\psi\| - \|F(\lambda)\psi\|)] = +\infty. \tag{4.3}$$

Proof. Necessity. If condition (4.3) is not satisfied for some nonzero element $\psi \in \mathfrak{N}_{\lambda_0}$, then there exists a sequence of numbers $\lambda_n \in \Pi_\epsilon$ converging to ∞ such that

$$\sup_n [|\lambda_n|(\|\psi\| - \|F(\lambda_n)\psi\|)] < \infty.$$

According to Theorem 2.2 and Observation 1 it follows that $\psi \in \mathfrak{N}_{\lambda_0,\infty}$, and therefore $\mathfrak{L}_\infty \neq D_A$. Thus we have proved the necessity of the condition.

Sufficiency. Let condition (4.3) be fulfilled for some nonzero element $\psi \in \mathfrak{N}_{\lambda_0}$. Consequently

$$\lim_{\substack{\lambda\to\infty \\ \lambda\in\Pi_\varepsilon}} \left[\frac{|\lambda|^2}{|\operatorname{Im}\lambda|}[\|\psi\| - \|F(\lambda)\psi\|)\right] = +\infty.$$

Then according to the Corollary to Theorem 4.4 the nonzero element ψ cannot belong to the linear manifold $\mathfrak{N}_{\lambda_0,\infty}$, so that $\mathfrak{N}_{\lambda_0,\infty}$ is the subspace consisting of zero only, and therefore $\mathfrak{L}_\infty = D_A$. This concludes the proof.*

Observation. The sufficient condition in Theorem 4.7 can be weakened if we replace the limit in (4.3) by the above limit. In the formulation of the necessary condition Π_ϵ can be replaced by Π in (4.3) if at the same time we replace $|\lambda|$ by $|\lambda|^2/|\operatorname{Im}\lambda|$.

---

* This theorem rectifies an error in [6]. There it is observed that operator functions $F(\lambda)$ ($\lambda \in \Pi$) corresponding to selfadjoint extensions of the second kind are characterized by the following condition: $\|F(\lambda)\psi\| < \|\psi\|$ for any nonzero $\psi \in \mathfrak{N}_{\lambda_0}$. In fact, this condition is only necessary but not sufficient in order that $F(\lambda)$ correspond to a selfadjoint operator of the second kind.

BIBLIOGRAPHY

[1] M. A. Naĭmark, *Self-adjoint extensions of the second kind of a symmetric operator*, Izv. Akad. Nauk SSSR Ser. Mat. 4 (1940), 53–104. (Russian) MR 2, 104.

[2] ———, *Spectral functions of a symmetric operator*, Izv. Akad. Nauk SSSR Ser. Mat. 4 (1940), 277–318. (Russian) MR 2, 105.

[3] A. V. Štraus, *On extensions of a symmetric operator depending on a parameter*, Izv. Akad. Nauk SSSR Ser. Mat. 29 (1965), 1389–1416; English transl., Amer. Math. Soc. Transl. (2) 61 (1967), 113–141. MR 34 #3338.

[4] ———, *Generalized resolvents of symmetric operators*, Izv. Akad. Nauk SSSR Ser. Mat. 18 (1954), 51–86. (Russian) MR 16, 48.

[5] C. Carathéodory, *Conformal representation*, Cambridge Univ. Press, New York, 1932; Russian transl., GITTL, Moscow, 1934.

[6] A. V. Štraus, *On generalized resolvents of a symmetric operator*, Dokl. Akad. Nauk SSSR 71 (1950), 241–244. (Russian) MR 11, 600.

Translated by:
H. N. Parker

Amer. Math. Soc. Transl.
(2) Vol. 90, 1970

AN EVOLUTIONARY EQUATION AND AN EXPANSION OF OPEN SYSTEMS*

UDC 519.3

Z. S. AGRANOVIČ

In this article we investigate open systems (in the sense of M. S. Livšic) in the case that the signals transformed by them have an arbitrary form and the corresponding operator groups can depend on time. We obtain an expansion of such a system in the form of a sequence of simpler systems.

Introduction

Open systems as mathematical models for physical systems interacting with the outside world were first introduced and studied in the recent works of M. S. Livšic [1], [2]. In these works he develops a theory of certain classes of open systems (important from the point of view of physical applications) in the steady-state condition, that is, when the waves (signals) being propagated in them are monochromatic and therefore dependence on time can be neglected. It is shown that such systems can be expanded under certain conditions into a sequence of elementary systems mutually bound together in a certain way. This result has obvious theoretical and practical interest.

In this article this result is generalized to the case when the signal which enters the system and is transformed by it and which is admissible by transmission channels connecting this system with the outside world can be of arbitrary form. The open system itself is not assumed to be invariant; its defining parameters (for example, capacitance and inductance in the case of a reactive multipole) may depend on time.

Following M. S. Livšic, we will study open systems for which the transmission channels are described by the equation[1)]

$$-i\frac{\partial\vec{\varphi}}{\partial t}=iBJ\frac{\partial\vec{\varphi}}{\partial x}\quad(x\neq 0),\tag{1}$$

---

* Translation of Izv. Akad. Nauk SSSR Ser. Mat. 30 (1966), 1201–1212.

1) The transmission channel equation can be reduced to this form in several important physical problems.

where $\vec{\phi}(x, t)$ is a vector function with values from the m-dimensional Euclidean space E^m $(m < \infty)$, B and J are linear operators on E^m such that $B > 0$, $J^* = J$ and $J^2 = I$, where I is the identity operator. Here we assume that for the semiaxis $x \leq 0$ the origin coincides with the input of the open system (F) and that for the semiaxis $x \geq 0$ the origin coincides with the output. This is represented schematically in Figure 1.

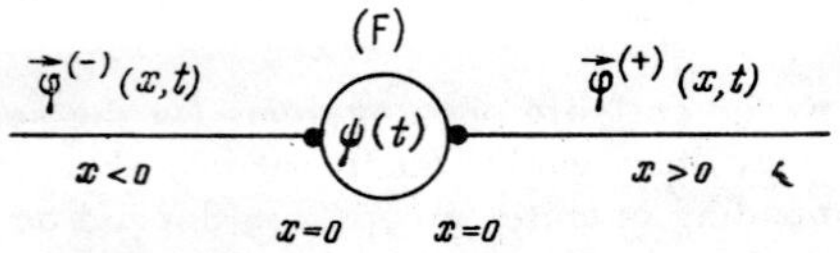

Figure 1

The condition of the transmission channels is characterized by the vector function

$$\vec{\varphi}(x, t) = \begin{cases} \vec{\varphi}^{(-)}(x, t), & x < 0, \\ \vec{\varphi}^{(+)}(x, t), & x > 0, \end{cases} \tag{2}$$

where $\vec{\phi}^{(-)}(x, t)$ and $\vec{\phi}^{(+)}(x, t)$ are solutions of equation (1) and are therefore uniquely determined[1)] by their values at $x = 0$. We will call the vector-functions $\vec{\phi}^{(-)}(t) = \vec{\phi}^{(-)}(0, t)$ and $\vec{\phi}^{(+)}(t) = \vec{\phi}^{(+)}(0, t)$ input and output signals respectively, and we will assume that the set of all signals forms a linear space $\widetilde{E}^m$, whose exact definition for the type of system under consideration will be given in §1.

We will describe the internal condition of the system (F) by the vector-function $\vec{\psi}(t)$ with values in the Hilbert space H; $\widetilde{H}$ will denote the linear space of all vector-functions $\vec{\psi}(t)$.

For the open system (F) we assume that the internal condition vector $\vec{\psi}(t)$ and the output signal $\vec{\phi}^{(+)}(t)$ depend only on the input signal $\vec{\phi}^{(-)}(t)$, that is,

$$\vec{\psi}(t) = R\vec{\varphi}^{(-)}(t), \quad \vec{\varphi}^{(+)}(t) = S\vec{\varphi}^{(-)}(t), \tag{3}$$

and we also suppose that the operators R and S are linear. Henceforth we will refer to the operator S as the transfer operator of the system (F).

---

1) An arbitrary solution of equation (1) has the form $\Phi(tI - xJB^{-1})\vec{\phi}_0$, where $\vec{\phi}_0$ is a constant vector from E^m and $\Phi(t)$ is a sufficiently smooth function.

Turning away now from obviously physical representations, we take the following general definition: the open linear system (F) is the collection of the two linear spaces $\widetilde{E}^m$ (space of signals) and $\widetilde{H}$ (space of internal conditions), the elements of which are vector-functions of t $(-\infty \leq t_1 \leq t \leq t_2 \leq \infty)$ with values in the Hilbert spaces E and H respectively, and of two linear operators: the operator R mapping $\widetilde{E}$ into $\widetilde{H}$, and the operator S mapping $\widetilde{E}$ into itself.

§1. Operators and operator groups corresponding to the open systems under consideration

We will find the operators R and S for the class of open systems which is of interest to us, the transmission channels of which are described by equation (1). Following a procedure almost identical to that of §2 of article [1], we get the following equations:

$$A\vec{\psi}(t) + \frac{i}{2} \sum_{\alpha, \beta=1}^{m} (\vec{\psi}(t), \vec{e}_\alpha)(J\vec{a}_\alpha, \vec{a}_\beta)\vec{e}_\beta + i\frac{d\vec{\psi}}{dt} = \sum_{\alpha=1}^{m} (\vec{\varphi}^{(-)}(t), \vec{a}_\alpha)\vec{e}_\alpha, \quad (4)$$

$$\varphi^{(+)}(t) = \vec{\varphi}^{(-)}(t) - iJ\sum_{\alpha=1}^{m} (\vec{\psi}(t), \vec{e}_\alpha)\vec{a}_\alpha, \quad (5)$$

where A is some Hermitian operator in the space H; $\vec{e}_\alpha$ $(\alpha = 1, 2, \cdots, m)$ are certain vectors from H, called channel vectors of system (F); and $\vec{a}_\alpha$ $(\alpha = 1, 2, \cdots, m)$ is an orthonormal basis in the space E^m.

We now introduce the operators (cf. [2]):

$$T\vec{\psi} = A\vec{\psi} + \frac{i}{2} \sum_{\alpha, \beta=1}^{m} (\vec{\psi}, \vec{e}_\alpha)(J\vec{a}_\alpha, \vec{a}_\beta)\vec{e}_\beta, \quad (6)$$

operating in the space H;

$$\Gamma\vec{\varphi} = \sum_{\alpha=1}^{m} (\vec{\varphi}, \vec{a}_\alpha)\vec{e}_\alpha, \quad (7)$$

mapping E^m into H; and

$$\Gamma^*\vec{\psi} = \sum_{\alpha=1}^{m} (\vec{\psi}, \vec{e}_\alpha)\vec{a}_\alpha, \quad (8)$$

mapping H into E^m. It is easy to see that Γ^* is an operator, adjoint to the operator Γ, and that

$$\frac{1}{i}(T - T^*) = \Gamma J \Gamma^*. \tag{9}$$

From (7) it also follows that

$$\Gamma \vec{a}_\alpha = \vec{e}_\alpha \quad (\alpha = 1, 2, \ldots, m). \tag{10}$$

Using the notation introduced above, we can rewrite equations (4) and (5) in the form

$$T\vec{\psi}(t) + i\frac{d\vec{\psi}}{dt} = \Gamma\vec{\varphi}^{(-)}(t), \tag{4'}$$

$$\vec{\varphi}^{(+)}(t) = \vec{\varphi}^{(-)}(t) - iJ\Gamma^*\vec{\psi}(t) \tag{5'}$$

and for the sake of greater generality we let the operators T, Γ, and Γ^* be functions of t:

$$T = T(t), \quad \Gamma = \Gamma(t), \quad \Gamma^* = \Gamma^*(t). \tag{11}$$

The physical significance of this last assumption is that the open system (i.e. the parameters that define it) can change in the course of time.

Equations (4')–(5'), from which we must find the operators R and S corresponding to the open system, may naturally be called the evolutionary equations of this open system.

In order to avoid unnecessary analytical difficulties we assume that for all values of t in the interval under consideration the operator $T(t)$ is bounded ($\Gamma(t)$ and $\Gamma^*(t)$ are obviously bounded) and that all operator-functions (11) in this interval are continuous in the uniform operator topology.[1)]

Given the initial condition $\vec{\psi}(t_0)$ at some point t_0 in the interval of continuity of the operator-function (11), there is a unique solution of equation (4') satisfying this condition. We can get this solution, by the method of variation of parameters, in the form

$$\vec{\psi}(t) = U_{t_0}^t(iT)\,\vec{\psi}(t_0) - i\int_{t_0}^{t} U_s^t(iT)\,\Gamma(s)\,\vec{\varphi}^{(-)}(s)\,ds, \tag{12}$$

where $U_{t_0}^t(iT)$ is an operator (called the evolutionary operator) defined by the conditions:

1) All the concepts of analysis that we use for operator-functions are contained, for example, in [4]. We point out that the concepts connected with the convergence of operators (derivative, integral, series) are henceforth considered with respect to the uniform operator topology.

$$\frac{d}{dt} U^t_{t_0}(iT) = iTU^t_{t_0}(iT), \quad U^{t_0}_{t_0}(iT) = I.$$

It is easy to see that $U^t_{t_0}(iT)$ can be represented by a series

$$U^t_{t_0}(iT) = I + \sum_{n=1}^{\infty} i^n \int_{t_0}^{t} T(\tau_1)\,d\tau_1 \int_{t_0}^{\tau_1} T(\tau_2)\,d\tau_2 \ldots \int_{t_0}^{\tau_{n-1}} T(\tau_n)\,d\tau_n, \tag{13}$$

which converges absolutely and uniformly with respect to t in each finite part of the interval of continuity of the operator $T(t)$, because it is majorized by the series

$$\sum_{n=0}^{\infty} \frac{h^n(t)}{n!}, \quad h(t) = \left| \int_{t_0}^{t} \| T(\tau) \| \, d\tau \right|.$$

Obviously if T does not depend on t, then

$$U^t_{t_0}(iT) = e^{i(t-t_0)T}. \tag{14}$$

The initial condition $\vec{\psi}(t_0)$ on the right side of formula (12) can be given in a different form, provided that it guarantees the linearity of the operator R (cf. the first of the formulas in (3)). In particular the following cases are of interest (we assume $t_0 = 0$, clearly without loss of generality):

1) $\vec{\psi}(0) = 0$ and consequently

$$\vec{\psi}(t) = R\vec{\varphi}^{(-)}(t) = -i\int_0^t U^t_s(iT)\,\Gamma(s)\,\vec{\varphi}^{(-)}(s)\,ds; \tag{15_1}$$

2) $\vec{\psi}(0) = (T_0 - \omega I)^{-1}\Gamma_0\vec{\phi}^{(-)}(0)$, where $T_0 = T(0)$, $\Gamma_0 = \Gamma(0)$, ω is a given number from the resolvent set of the operator R, and hence

$$\vec{\psi}(t) = R\vec{\varphi}^{(-)}(t)$$

$$= U^t_0(iT)(T_0 - \omega I)^{-1}\Gamma_0\vec{\varphi}^{(-)}(0) - i\int_0^t U^t_s(iT)\,\Gamma(s)\,\vec{\varphi}^{(-)}(s)\,ds. \tag{15_2}$$

The physical significance of formula (15_1) is clearly that the corresponding open system is at rest (in an unexcited state) up until $t = 0$. Formula (15_2) can be interpreted in the following manner: up until $t = 0$ the open system was invariant and was operating in the steady state condition with frequency ω, and then arbitrary signals began to impinge upon it and the system itself began to change with time. We observe that formula (15_2)

includes the steady-state condition, because assuming $T(t) \equiv T_0$ and $\Gamma(t) \equiv \Gamma_0$, and taking $\vec{\phi}^{(-)}(t) = \vec{\phi}_0^{(-)} \cdot e^{i\omega t}$, we get for (15_2)

$$\vec{\psi}(t) = \vec{\psi}_0 \cdot e^{i\omega t}, \quad (T_0 - \omega I)\vec{\psi}_0 = \Gamma_0 \vec{\varphi}_0^{(-)},$$

that is, we get the basic equation of M. S. Livšic [1].

Having the operator R, directly from (5′) we find the transfer operator:

$$S = I - iJ\Gamma^* R. \tag{16}$$

Henceforth we limit ourselves to the investigation of open systems for which the corresponding operators R and S are given by formulas (15) and (16). For the space of signals $\widetilde{E}^m$ and the space of inner conditions $\widetilde{H}$ we will take spaces of vector-functions of t with values respectively in E^m and H which are strongly piecewise continuous in the interval under consideration.

Just as in the case of systems in the steady-state condition, the concept of operator group (cf. [2]) will play a crucial role in the following: Recall that the operator group[1)] $[T, H; J, E^m; \Gamma]$ is the set consisting of the operator T in the space H, the operator J in the space E^m and having the property $J^* = J$ and $J^2 = 1$, and the operator Γ from E^m to H, all connected by relation (9).

According to formulas (15)–(16), to each open system of the type under consideration there corresponds (belongs) an operator group $[T(t), H; J, E^m; \Gamma(t)]$, where the operators T and Γ can depend on t. Conversely, given such an operator group $[T(t), H; J, E^m; \Gamma(t)]$, by (15)–(16) we can construct an open system of the form under consideration having as its operator group the one given.

§2. The coupling of operators and operator groups

The concept of the coupling of operators and operator groups is one of the fundamental concepts lying at the base of the subsequent investigation (cf. [2]). Recall the following definitions.

Let the Hilbert space H be the orthogonal sum of its subspaces H_1 and H_2; let T_1 and T_2 be operators mapping H_1 and H_2 respectively into themselves; let C be any operator mapping H_1 into H_2 and mapping H_2 into 0. Then the operator T mapping H into H and given by the formula

1) For convenience in printing we use a different notation for the operator group than that employed in [2].

$$T = T_1P_1 + T_2P_2 + C,$$

where P_k is a projection operator on H_k $(k = 1, 2)$, is called the coupling of T_2 with T_1 and is designated by $T = T_2 \curlyvee T_1$. C is called the coupling operator. Clearly the space H_2 is invariant relative to $T = T_2 \curlyvee T_1$.

The coupling of the operator group $[T_2, H_2; J, E^m; \Gamma_2]$ with the group $[T_1, H_1; J, E^m, \Gamma_1]$ is the operator group

$$[T, H; J, E^m; \Gamma] \equiv [T_2, H_2; J, E^m; \Gamma_2] \quad \curlyvee [T_1, H_1; J, E^m; \Gamma_1],$$

where $T = T_2 \curlyvee T_1$ with the operator of coupling $C = iP_2\Gamma_2 J\Gamma_1^* P_1$, $\Gamma = \Gamma_1 + \Gamma_2$. That this definition is correct, i.e. that equation (9) is correct for T, J and Γ, is immediately verifiable.

Lemma 1. *Let $[T, H; J, E^m; \Gamma]$ be some operator group. If $H_2 \subset H$ is an invariant subspace for T, if $H_1 = H \ominus H_2$, and if the P_k are projection operators on H_k $(k = 1, 2)$, then $[T_k, H_k; J, E^m; \Gamma_k]$, where $T_k = P_kT$, $\Gamma_k = P_k\Gamma$ $(k = 1, 2)$, is an operator group and*

$$[T, H; J, E^m; \Gamma] = [T_2, H_2; J, E^m; \Gamma_2] \curlyvee [T_1, H_1; J, E^m; \Gamma_1).$$

There is a proof of this lemma in [1] (applicable to operator complexes).

If

$$T = T_2 \curlyvee T_1 = T_1P_1 + T_2P_2 + C, \tag{17}$$

then, as is known (cf. [3]),

$$\begin{aligned}(T - \omega I)^{-1} & \\ = \ (T_1 - \omega I)^{-1}P_1 + (T_2 - \omega I)^{-1}P_2 &- (T_2 - \omega I)^{-1}C(T_1 - \omega I)^{-1}P_1,\end{aligned} \tag{18}$$

where ω is a point at which the operators T_1 and T_2 are regular (from this it follows that the operator $T = T_2 \curlyvee T_1$ is regular at ω).

We still need an analog of formula (18) for the operator $U_s^t(iT)$.

Lemma 2. *Let the Hilbert space H be the orthogonal sum of its subspaces H_1 and H_2; let the vector-functions $T(t)$, $T_1(t)$ and $T_2(t)$ mapping H, H_1 and H_2 respectively into themselves be continuous in the interval $[t_1, t_2]$, and for all t in this interval let them be related by the equation*

$$T(t) = T_2(t) \curlyvee T_1(t) = T_1(t)P_1 + T_2(t)P_2 + C(t). \tag{19}$$

Then for all $s, t \in [t_1, t_2]$

$$U_s^t(iT) = U_s^t(iT_1)P_1 + U_s^t(iT_2)P_2 + i\int_s^t U_\tau^t(iT_2)C(\tau)U_s^\tau(iT_1)\,d\tau\cdot P_1. \tag{20}$$

Proof. Using (19) and the equation

$$C(t)P_2 = P_1C(t) = 0, \quad P_2C(t) = C(t)P_1 = C(t),$$

we find

$$\int_s^t T(\tau)\,d\tau = \int_s^t T_1(\tau)\,d\tau\cdot P_1 + \int_s^t T_2(\tau)\,d\tau\cdot P_2 + \int_s^t C(\tau)\,d\tau\cdot P_1,$$

$$\int_s^t T(\tau_1)\,d\tau_1\int_s^{\tau_1} T(\tau_2)\,d\tau_2 = \int_s^t T_1(\tau_1)\,d\tau_1\int_s^{\tau_1} T_1(\tau_2)\,d\tau_2\cdot P_1$$

$$+ \int_s^t T_2(\tau_1)\,d\tau_1\int_s^{\tau_1} T_2(\tau_2)\,d\tau_2\cdot P_2$$

$$+ \left[\int_s^t T_2(\tau_1)\,d\tau_1\int_s^{\tau_1} C(\tau_2)\,d\tau_2 + \int_s^t C(\tau_1)\,d\tau_1\int_s^{\tau_1} T_1(\tau_2)\,d\tau_2\right]\cdot P_1,$$

$$\cdots\cdots\cdots\cdots\cdots\cdots$$

and then, by induction,

$$\int_s^t T(\tau_1)\,d\tau_1\int_s^{\tau_1} T(\tau_2)\,d\tau_2\ldots\int_s^{\tau_{n-1}} T(\tau_n)\,d\tau_n = \int_s^t T_1(\tau_1)\,d\tau_1\ldots\int_s^{\tau_{n-1}} T_1(\tau_n)\,d\tau_n\cdot P_1$$

$$+ \int_s^t T_2(\tau_1)\,d\tau_1\ldots\int_s^{\tau_{n-1}} T_2(\tau_n)\,d\tau_n\cdot P_2 + A_n(s,t)\cdot P_1, \tag{21}$$

where

$$A_n(s,t) = \sum_{k=0}^{n-1}\int_s^t T_2(\tau_1)\,d\tau_1\ldots$$

$$\ldots\int_s^{\tau_{k-1}} T_2(\tau_k)\,d\tau_k\int_s^{\tau_k} C(\tau_{k+1})\,d\tau_{k+1}\int_s^{\tau_{k+1}} T_1(\tau_{k+2})\,d\tau_{k+2}\ldots\int_s^{\tau_{n-1}} T_1(\tau_n)\,d\tau_n$$

$$(\tau_0 = t;\quad n = 1, 2, \ldots).$$

Turning now to formula (13), by virtue of (21) and the equation $I = P_1 + P_2$ we get

$$U_s^t\ (iT) = U_s^t\ (iT_1)P_1 + U_s^t\ (iT_2)P_2 + \sum_{n=1}^{\infty} i^n A_n(s,t).$$

Also note that by virtue of (13)

$$i\int_s^t U_\tau^t\,(iT_2)C(\tau)U_s\,(iT_1)\,d\tau = i\int_s^t \Big\{I + \sum_{k=1}^{\infty} i^k \int_\tau^t T_2(\tau_1)\,d\tau_1 \dots \int_\tau^{\tau_{k-1}} T_2(\tau_k)\,d\tau_k\Big\}$$

$$\times\, C(\tau)\Big\{I + \sum_{m=1}^{\infty} i^m \int_s^\tau T_1(\sigma_1)\,d\sigma_1 \dots \int_s^{\sigma_{m-1}} T_1(\sigma_m)\,d\sigma_m\Big\}d\tau$$

$$= \sum_{n=1} i^n \sum_{k=0}^{n-1} \int_s^t \Big\{\int_\tau^t T_2(\tau_1)\,d\tau_1 \dots \int_\tau^{\tau_{k-1}} T_2(\tau_k)\,d\tau_k\Big\} C(\tau)\Big\{\int_s^\tau T_1(\sigma_1)\,d\sigma_1 \dots$$

$$\dots \int_s^{\sigma_{n-k-2}} T_1(\sigma_{n-k-1})\,d\sigma_{n-k-1}\Big\}d\tau = \sum_{n=1}^{\infty} i^n \sum_{k=0}^{n-1} \int_s^t T_2(\tau_1)\,d\tau_1 \dots$$

$$\dots \int_s^{\tau_{k-1}} T_2(\tau_k)\,d\tau_k \int_s^{\tau_k} C(\tau_{k+1})\,d\tau_{k+1} \int_s^{\tau_{k+1}} T_1(\tau_{k+2})\,d\tau_{k+2} \dots$$

$$\dots \int_s^{\tau_{n-1}} T_1(\tau_n)\,d\tau_n = \sum_{n=1}^{\infty} i^n A_n(s,t),$$

Thus we arrive at formula (20).

Corollary. *If the operators* T, T_1 *and* T_2 *do not depend on* t *and are related by equation* (17), *then*

$$e^{itT} = e^{itT_1}P_1 + e^{itT_2}P_2 + i\int_0^t e^{i(t-\tau)T_2}Ce^{i\tau T_1}\,d\tau\cdot P_1. \qquad (22)$$

This formula follows immediately from (14) and (20).

§3. Expansion of an open system into a sequence of simpler systems

We now turn to the basic result in the representation of open systems of the type under consideration in the form of a chainwise union of simpler open systems. The concept of chainwise union is introduced in (1) and is defined in the following manner.

Let (F_k), $k = 1, 2, \cdots, n$, be open systems to which correspond spaces of signals $\widetilde{E}^m_k$, spaces of internal conditions $\widetilde{H}_k$, and operators R_k and S_k; moreover assume that all the $\widetilde{E}^m_k$ coincide, i.e. that $\widetilde{E}^m_k = \widetilde{E}^m$ for $k = 1, 2, \cdots, n$. The open system (F) to which correspond the space of signals $\widetilde{E}^m$, the space of internal conditions

$$\widetilde{H} = \widetilde{H}_1 \oplus \widetilde{H}_2 \oplus \cdots \oplus \widetilde{H}_n$$

and the operators

$$R = R_1 + R_2S_1 + \ldots + R_nS_{n-1} \ldots S_2S_1,$$
$$S = S_nS_{n-1} \ldots S_2S_1,$$

is called the chainwise union of the systems $(F_1), (F_2), \cdots, (F_n)$ and is denoted by

$$(F) = (F_1) \curlyvee (F_2) \curlyvee \ldots \curlyvee (F_n).$$

The physical significance of the chainwise union is that the output of each system (F_k) is connected with the input of the next system (F_{k+1}) and with the outside world by means of the transmission channels. This is represented schematically in Figure 2.

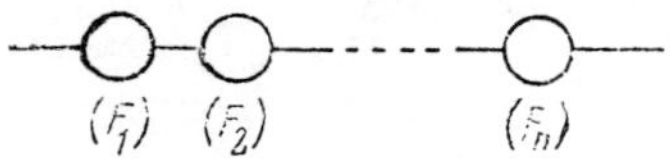

Figure 2

Theorem 1. *If the operator group* $[T(t), H; J, E^m; \Gamma(t)]$ *belonging to the open system* (F) *and defined by formulas* (15)–(16) *is for all values of* t *under consideration the coupling*

$$[T(t), H; J, E^m; \Gamma(t)]$$
$$= [T_2(t), H_2; J, E^m; \Gamma_2(t)] \curlyvee [T_1(t), H_1; J, E^m; \Gamma_1(t)],$$

where H_1 *and* H_2 *do not depend on* t, *then*

$$(F) = (F_1) \curlyvee (F_2),$$

where (F_k) *is the open system to which the operator group* $[T_k(t), H_k; J, E^m; \Gamma_k(t)]$ *belongs for* $k = 1, 2$.

Proof. Let the spaces $\widetilde{E}^m$ and $\widetilde{H}$ and the operators R and S correspond to the system (F); let E_k^m, H_k, R_k, and S_k correspond to (F_k), $k = 1, 2$. It follows immediately that $\widetilde{E}_k^m = \widetilde{E}^m$, $k = 1, 2$, and that $\widetilde{H} = \widetilde{H}_1 \oplus \widetilde{H}_2$. Thus it remains to show

$$R = R_1 + R_2 S_1, \quad S = S_2 S_1.$$

First consider formula (15_1). Using (20) and the equation

$$\Gamma(t) = \Gamma_1(t) + \Gamma_2(t),$$

we have for any $\vec{\phi}^{(-)}(t) \in \widetilde{E}^m$:

$$R\vec{\varphi}^{(-)}(t) = -i\int_0^t U_s^t(iT)\,\Gamma(s)\,\vec{\varphi}^{(-)}(s)\,ds = -i\int_0^t \Big\{U_s^t(iT_1)\,P_1 + U_s^t(iT_2)\,P_2$$

$$+ i\int_s^t U_\tau^t(iT_2)\,C(\tau)\,U_s^\tau(iT_1)\,d\tau\cdot P_1\Big\}\{\Gamma_1(s) + \Gamma_2(s)\}\,\vec{\varphi}^{(-)}(s)\,ds,$$

and because $\Gamma_k(t) = P_k\Gamma_k(t)$, $k = 1, 2$, $C(t) = iP_2\Gamma_2(t)J\Gamma_1^*(t)P_1$, it follows that

$$R\vec{\varphi}^{(-)}(t) = -i\int_0^t U_s^t(iT_1)\,\Gamma_1(s)\,\vec{\varphi}^{(-)}(s)\,ds - i\int_0^t U_s^t(iT_2)\,\Gamma_2(s)\,\vec{\varphi}^{(-)}(s)\,ds$$

$$+ i\int_0^t\Big\{\int_s^t U_\tau^t(iT_2)\,\Gamma_2(\tau)\,J\Gamma_1^*(\tau)\,U_s^\tau(iT_1)\,d\tau\Big\}\,\Gamma_1(s)\,\vec{\varphi}^{(-)}(s)\,ds$$

$$= R_1\vec{\varphi}^{(-)}(t) - i\int_0^t U_s^t(iT_2)\,\Gamma_2(s)\,\vec{\varphi}^{(-)}(s)\,ds$$

$$+ i\int_0^t U_\tau^t(iT_2)\,\Gamma_2(\tau)\,J\Gamma_1^*(\tau)\,d\tau\int_0^\tau U_s^\tau(iT_1)\,\Gamma_1(s)\,\vec{\varphi}^{(-)}(s)\,ds$$

$$= R_1\vec{\varphi}^{(-)}(t)$$

$$-i\int_0^t U_s^t(iT_2)\,\Gamma_2(s)\Big\{\vec{\varphi}^{(-)}(s) - J\Gamma_1^*(s)\int_0^s U_\tau^s(iT_1)\,\Gamma_1(\tau)\,\vec{\varphi}^{(-)}(\tau)\,d\tau\Big\}\,ds.$$

Since by formulas (15_1) and (16)

$$\vec{\varphi}^{(-)}(s) - J\Gamma_1^*(s)\int_0^s U_\tau^s(iT_1)\,\Gamma_1(\tau)\,\vec{\varphi}^{(-)}(\tau)\,d\tau$$
$$= \vec{\varphi}^{(-)}(s) - iJ\Gamma_1^*(s)\,R_1\vec{\varphi}^{(-)}(s) = S_1\vec{\varphi}^{(-)}(s),$$

we get

$$R\vec{\varphi}^{(-)}(t) = R_1\vec{\varphi}^{(-)}(t) + R_2S_1\vec{\varphi}^{(-)}(t),$$

and hence, since $\vec{\phi}^{(-)}(t) \in \widetilde{E}^m$ is arbitrary,

$$R = R_1 + R_2S_1.$$

Now let R be defined by formula (15_2). By what was proved above

$$\begin{aligned} -i\int_0^t U_s^t(iT)\,\Gamma(s)\,\vec{\varphi}^{(-)}(s)\,ds &= -i\int_0^t U_s^t(iT_1)\,\Gamma_1(s)\,\vec{\varphi}^{(-)}(s)\,ds \\ -i\int_0^t U_s^t(iT_2)\,\Gamma_2(s)&\left\{\vec{\varphi}^{(-)}(s) - J\Gamma_1^*(s)\int_0^s U_\tau^s(iT_1)\,\Gamma_1(\tau)\,\vec{\varphi}^{(-)}(\tau)\,d\tau\right\}ds. \end{aligned} \tag{23}$$

Using formulas (18) and (20) we find further that

$$U_0^t(iT)\,(T(0) - \omega I)^{-1}\,\Gamma(0)\,\vec{\varphi}^{(-)}(0)$$
$$= \left\{U_0^t(iT_1)\,P_1 + U_0^t(iT_2)\,P_2 + i\int_0^t U_s^t(iT_2)\,C(s)\,U_0^s(iT_1)\,ds\cdot P_1\right\}$$
$$\times\{(T_1(0) - \omega I)^{-1}P_1 + (T_2(0) - \omega I)^{-1}P_2 - (T_2(0) - \omega I)^{-1}C(0)(T_1(0) - \omega I)^{-1}P_1\}$$
$$\times\{\Gamma_1(0) + \Gamma_2(0)\}\,\vec{\varphi}^{(-)}(0) = U_0^t(iT_1)\,(T_1(0) - \omega I)^{-1}\,\Gamma_1(0)\,\vec{\varphi}^{(-)}(0)$$
$$+ U_0^t(iT_2)\,(T_2(0) - \omega I)^{-1}\,\Gamma_2(0)\,\{I - iJ\Gamma_1^*(0)\,(T_1(0) - \omega I)^{-1}\,\Gamma_1(0)\}\,\vec{\varphi}^{(-)}(0)$$
$$-\int_0^t U_s^t(iT_2)\,\Gamma_2(s)\,J\Gamma_1^*(s)\,U_0^s(iT_1)\,ds\,(T_1(0) - \omega I)^{-1}\,\Gamma_1(0)\,\vec{\varphi}^{(-)}(0).$$

Combining this equation with the preceding one, we get

$$R\vec{\varphi}^{(-)}(t) = R_1\vec{\varphi}^{(-)}(t) + U_0^t(iT_2)(T_2(0) - \omega I)^{-1}\Gamma_2(0)\,\{I - iJ\Gamma_1^*(0)\,(T_1(0)$$
$$- \omega I)^{-1}\,\Gamma_1(0)\}\,\vec{\varphi}^{(-)}(0) - i\int_0^t U_s^t(iT_2)\,\Gamma_2(s)\,\Big\{\vec{\varphi}^{(-)}(s)$$
$$- iJ\Gamma_1^*(s)\Big[U_0^s(iT_1)\,(T_1(0) - \omega I)^{-1}\,\Gamma_1(0)\,\vec{\varphi}^{(-)}(0) -$$

$$- i \int_0^s U_\tau^s (iT_1) \Gamma_1(\tau) \vec{\varphi}^{(-)}(\tau)\, d\tau \Big]\Big\} ds = R_1 \vec{\varphi}^{(-)}(t) + R_2 S_1 \vec{\varphi}^{(-)}(t).$$

Finally, by formula (16) we have for the transfer operator S that

$$S = I - iJ\Gamma^* R = I - iJ(\Gamma_1^* + \Gamma_2^*)(R_1 + R_2 S_1)$$

$$= I - iJ\Gamma_1^* R - iJ\Gamma_2^* R_2 S_1 = S_1 - iJ\Gamma_2^* R_2 S_1 = (I - iJ\Gamma_2^* R_2) S_1 = S_2 S_1$$

(because $\Gamma_k^* = \Gamma_k^* P_k$, $k = 1, 2$), and the theorem is proved.

Observation 1. In Theorem 1, as was stipulated earlier, it is assumed that the operator-functions $T(t)$ and $\Gamma(t)$ are continuous in the interval under consideration. The theorem is valid and can be proved in an analogous way in the case that the operator-functions $T(t)$ and $\Gamma(t)$ are piecewise continuous (in particular, piecewise constant) provided that they are normalized at each point t_0 of discontinuity by the condition

$$T(t_0) = T(t_0 - 0), \quad \Gamma(t_0) = \Gamma(t_0 - 0).$$

Observation 2. In applications (for example, in the case of a reactive multipole with variable parameters) it may happen that not only the operators T and Γ, belonging to the open system (F) and contained in the operator group, but also the scalar product in H change with time. Theorem 1 remains valid even in this case if only (as is indicated in the conditions of the theorem) the orthogonal subspaces H_1 and H_2 do not depend on t (i.e. if the projection operators P_1 and P_2 do not depend on t).

From Lemma 1 and the theorem just proved we get immediately

Corollary. *Let* (F) *be an open system defined by* (15)–(16), *let* $[T(t), H; J, E^m; \Gamma(t)]$ *be the operator group that belongs to it, let*

$$H = \mathfrak{H}_0 \supset \mathfrak{H}_1 \supset \mathfrak{H}_2 \supset \ldots \supset \mathfrak{H}_{n-1} \supset 0$$

be a chain of subspaces which are invariant relative to $T(t)$ *for all values of* t *under consideration, and let* $H_k = \mathfrak{H}_{k-1} \ominus \mathfrak{H}_k$ $(k = 1, 2, \cdots, n)$ *not depend on* t *(i.e. the scalar product in* H*, although it may vary with* t*, is such that the orthogonality of its elements is still preserved).*

Then $(F) = (F_1) \curlyvee (F_2) \curlyvee \cdots \curlyvee (F_n)$, *where* (F_k) *is an open system of the form defined by* (15)–(16) *to which belongs the operator group* $[T_k(t), H_k; J, E^m; \Gamma_k(t)]$ *with* $T_k(t) = P_k T(t)$, $\Gamma_k(t) = P_k \Gamma(t)$ $(P_k$ *is the projection operator on* $H_k)$.

Hence it follows that if the operator group belonging to the open system

(F) does not depend on t and the space H is finite dimensional, then (F) can be represented as a chainwise union of elementary open systems (F_k) such that each H_k is one dimensional.

In conclusion we investigate an open system such that the operator group $[T, H; J, E^m; \Gamma]$ does not depend on t, $\dim H = 1$, and hence T is a product operator on some number τ. Supposing that up until $t = 0$ this system is at rest, we find by (14) and (15_1) that

$$\vec{\psi}(t) = R\vec{\varphi}^{(-)}(t) = -i\int_0^t e^{i(t-s)\tau}\,\Gamma\vec{\varphi}^{(-)}(s)\,ds.$$

Let $\vec{a}_\alpha$ $(\alpha = 1, 2, \cdots, m)$ be an orthonormal basis in E^m, and let $\vec{\psi}_0$ be the unit vector in H. Letting

$$(\vec{\varphi}^{(-)}(s), \vec{a}_\alpha) = \varphi_\alpha^{(-)}(s), \quad (\Gamma\vec{a}_\alpha, \vec{\psi}_0) = \varepsilon_\alpha, \tag{24}$$

so that $\Gamma\vec{a}_\alpha = \epsilon_\alpha\vec{\psi}_0$, we have

$$\Gamma\vec{\varphi}^{(-)}(s) = \sum_{\alpha=1}^m \varphi_\alpha^{(-)}(s)\,\Gamma\vec{a}_\alpha = \sum_{\alpha=1}^m \varepsilon_\alpha\varphi_\alpha^{(-)}(s)\,\vec{\psi}_0$$

and hence

$$\vec{\psi}(t) = R\vec{\varphi}^{(-)}(t) = -ie^{-it\tau}\sum_{\alpha=1}^m \varepsilon_\alpha\int_0^t e^{-is\tau}\varphi_\alpha^{(-)}(s)\,ds\,\vec{\psi}_0. \tag{25}$$

Substituting (25) into formula (16) and using the equation

$$\Gamma^*\vec{\psi}_0 = \sum_{\beta=1}^m (\Gamma^*\vec{\psi}_0, \vec{a}_\beta)\,\vec{a}_\beta = \sum_{\beta=1}^m (\vec{\psi}_0, \Gamma\vec{a}_\beta)\,\vec{a}_\beta = \sum_{\beta=1}^m \bar{\varepsilon}_\beta\vec{a}_\beta \tag{26}$$

(cf. (24)), we find the output signal of the elementary system under consideration in the form

$$\vec{\varphi}^{(+)}(t) = S\vec{\varphi}^{(-)}(t) = \vec{\varphi}^{(-)}(t) - e^{-it\tau}\sum_{\alpha=1}^m \varepsilon_\alpha\int_0^t e^{-is\tau}\varphi_\alpha^{(-)}(s)\,ds\sum_{\beta,\gamma=1}^m \bar{\varepsilon}_\beta J_{\beta\gamma}\vec{a}_\gamma, \tag{27}$$

where

$$J_{\beta\gamma} = (J\vec{a}_\beta, \vec{a}_\gamma). \tag{28}$$

Notice further that condition (9) for the operator group establishes a relationship between the number τ and the numbers ϵ_α and $J_{\alpha\beta}$. In order to find this relationship we investigate the equation

$$(-i(T - T^*)\vec{\psi}_0, \vec{\psi}_0) = (\Gamma J\Gamma^*\vec{\psi}_0, \vec{\psi}_0)$$

which follows from (9). By virtue of (26) and (28)

$$(\Gamma J\Gamma^*\vec{\psi}_0, \vec{\psi}_0) = (J\Gamma^*\vec{\psi}_0, \Gamma^*\vec{\psi}_0) = \sum_{\alpha,\beta=1}^{m} J_{\alpha\beta}\bar{\varepsilon}_\alpha\varepsilon_\beta,$$

Because of this and the fact that $(-i(T - T^*)\vec{\psi}_0, \vec{\psi}_0) = 2\,\mathrm{Im}\,\tau$, we have

$$2\mathrm{Im}\,\tau = \sum_{\alpha,\beta=1}^{m} J_{\alpha\beta}\bar{\varepsilon}_\alpha\varepsilon_\beta. \tag{29}$$

BIBLIOGRAPHY

[1] M. S. Livšic, *On linear physical systems connected with the outside world by transmission channels*, Izv. Akad. Nauk SSSR Ser. Mat. 27 (1963), 993–1030. (Russian)

[2] ———, *Open systems as linear automata*, Izv. Akad. Nauk SSSR Ser. Mat. 27 (1963), 1215–1228. (Russian) MR **31** #6739.

[3] M. S. Brodskiĭ and M. S. Livšic, *Spectral analysis of non-selfadjoint operators and intermediate systems*, Uspehi Mat. Nauk **13** (1958), no. 1 (79), 3–85; English transl., Amer. Math. Soc. Transl. (2) **13** (1960), 265–346. MR 20 #7221; 22 #3982.

[4] E. Hille and R. Phillips, *Functional analysis and semi-groups*, rev. ed., Amer. Math. Soc. Colloq. Publ., vol. 31, Amer. Math. Soc., Providence, R. I., 1957; Russian transl., IL, Moscow, 1962. MR **19**, 664.

Translated by:
H. Neal Parker

Amer. Math. Soc. Transl.
(2) Vol. 90, 1970

ANALYTIC PROBLEMS AND RESULTS IN THE THEORY OF LINEAR OPERATORS IN HILBERT SPACE*

M. G. KREĬN

This is the sixtieth year since David Hilbert's fourth article appeared in the Göttingen Nachrichten. In that article, Hilbert summarized his research in the theory of quadratic forms in an infinite number of variables and first obtained a spectral resolution for an arbitrary bounded selfadjoint operator.

Hilbert's work determined the development of operator theory for many years; this was one impetus of many for which modern mathematics is obliged to this great scholar.

Hilbert's research immediately drew the attention of many prominent mathematicians the world over; at the same time, it was only towards the end of the twenties that J. von Neumann and M. Stone worked out the precise concept of an unbounded selfadjoint operator and, in a generalization of Hilbert's result, obtained a spectral resolution for such an operator. Simultaneously, von Neumann constructed an exceptionally transparent theory of the extension of Hermitian operators.

By that time, something unforeseen and improbable had already happened. In the past, when mechanics and physics resorted to linear differential equations to describe some phenomenon, they did so with the deep conviction that the true equations were nonlinear and that their investigations were first approximations.

Suddenly, the new mechanics—quantum mechanics—announced that linearity was the law of nature for operators describing the microworld. Moreover, these operators turned out to be exactly the operators acting in Hilbert space, the space of states of a micro-object.

Henceforth, physical intuition, with its boldness and insight, was joined to the efforts of the mathematical mind. This is not lightly said. In connection with various problems in quantum statistics and quantum field theory, a set of half-finished products (from the mathematical point of view) has

*Translation of Proc. Internat. Congress Math. (Moscow, 1966), "Mir", Moscow, 1968, pp. 189–216.

accumulated up to the present time in the perturbation theory of linear operators and it is difficult even approximately to picture them floating in the mathematical crucible. For example, take the concept of the Heisenberg S-matrix, which arises repeatedly in mathematics. This concept found its precise mathematical form in the perturbation theory of selfadjoint operators: the S-matrix is found in classical problems in mathematical physics, in problems related to the theory of stationary random processes and, most strikingly, in various problems in the theory of nonselfadjoint operators.

Very significant results in Hilbert space theory are obtained in the process of tending to the purely internal development of the theory. At the present time we observe that events are moving fast in the arena of Hilbert spaces, indicating that operator theory in these spaces is full of life and turning to great new problems.

I can speak here of only one chain of events. I do not know if I will succeed in transmitting this, but all these events belong to some coherent set to some ensemble with a quite singular architecture, with its own analytic appartus and, one can even say, with its own calculus. Here are various problems in the theory of Hermitian, selfadjoint and nonselfadjoint operators, in perturbation theory, in scattering theory, diverse problems in the theory of spaces with indefinite metric, the theory of direct and inverse spectral problems for canonical equations, various analytic problems and many others.

§1. Certain aspects of perturbation theory

Let us agree on some notation and terminology. Let $\mathfrak{R} = \mathfrak{R}(\mathfrak{H})$ denote the Banach algebra of all bounded linear operators on a separable Hilbert space $\mathfrak{H}$ and let $\mathfrak{S}_\infty$ denote the closed ideal, in the ring $\mathfrak{R}$, consisting of all completely continuous operators.

To each operator $A \in \mathfrak{S}_\infty$ there corresponds a sequence of s-numbers $\{s_j(A)\}_1^\infty$, defined as the sequence of all eigenvalues of the nonnegative operator $|A| = (A^*A)^{1/2} \geq 0$, numbered (with multiplicity) in decreasing order.

Let $\mathfrak{S}_p$ denote the von Neumann-Schatten ideals:

$$\mathfrak{S}_p = \{A\colon\ A \in \mathfrak{S}_\infty,\quad \sum s_j^p(A) < \infty\}.$$

The ideal $\mathfrak{S}_1$ of nuclear operators plays a fundamental role. Among the operators in $\mathfrak{R}$, a nuclear operator A is characterized by the fact that it has an absolutely convergent matrix trace in any orthonormal basis $\{e_j\}_1^\infty$, i.e. the series $\Sigma_{j=1}^\infty (Ae_j, e_j)$ converges absolutely. This matrix trace, by a theorem of V. B. Lidskiĭ, coincides with the spectral trace, i.e. with the sum of the eigenvalues of A:

$$\operatorname{sp} A = \sum (Ae_j, e_j) = \sum \lambda_j (A).$$

For a nuclear operator A, the quantity

$$\det (I - A) = \lim \det \| \delta_{jk} - (Ae_j, e_k) \|_1^n = \prod_j (1 - \lambda_j (A))$$

is well defined.

It is well known that in perturbation theory, the nuclear operators are distinguished from the other completely continuous operators by their comparatively quiet character. We have in mind the following: let H_0 and H_1 be two selfadjoint operators differing by a nuclear operator, $H_1 = H_0 + V$ ($V \in \mathfrak{S}_1$). Then the absolutely continuous spectra of H_0 and H_1 are always the same. Moreover, the absolutely continuous parts of these operators are unitarily equivalent. This was brought out as a result of the work of Rosenblum and Kato. As far as these are concerned, as well as a series of other preceding and succeeding important investigations by M. Š. Birman, T. Kato, J. Cook, S. Kuroda, O. A. Ladyženskaya, C. Møller, P. A. Rejto, L. D. Faddeev, K. O. Friedrichs (fundamental), J. M. Jauch and others, I can only refer to two books on perturbation theory—those of Friedrichs [1] and Kato [2].

As H. Weyl, von Neumann and Kuroda established, we can always reduce the continuous spectrum to a point by means of perturbations V from ideals larger than $\mathfrak{S}_1$. We will now indicate an ideal of completely continuous operators which contains each ideal $\mathfrak{S}_p$ and all of whose operators turn out to be completely trustworthy for other problems in perturbation theory. This remarkable ideal, denoted by $\mathfrak{S}_\omega$, was introduced by V. I. Macaev and is defined by:

$$\mathfrak{S}_\omega = \left\{ A: \ A \in \mathfrak{S}_\infty, \ \sum \frac{1}{n} s_n (A) < \infty \right\} .$$

It turns out that many of the definitive statements in the spectral analysis of nonselfadjoint operators can be stated solely in terms of this ideal.

As an example, we give the following theorem of Macaev [3a]:

Suppose the spectrum of the operator $H = H^*$ is discrete (i.e. suppose the spectrum consists of eigenvalues of finite multiplicity with a single *condensation point at infinity). Then for each* $V \in \mathfrak{S}_\omega$, *the system of root vectors of the operator* $A = H + V$ *is complete in* $\mathfrak{H}$.

If a completely continuous operator V *is not in* $\mathfrak{S}_\omega$ *then there is a selfadjoint operator* H *with discrete spectrum such that the spectrum of* $A = H + V$ *is not complete.*

The first part of this assertion may be strengthened so as to supplement a theorem of M. V. Keldyš [4] on completeness.

Generally speaking, completeness problems such as these for root vectors of linear operators are not part of the scheme of this report. In particular, much research by (mainly) Soviet mathematicians that is based on Keldyš's fundamental work [4] falls outside this scheme. Some idea of the possibilities and achievements here can be obtained from Gohberg and Kreĭn's first book [6a] and from the survey report by Keldyš and Lidskiĭ [5].

In studies of the spectral properties of operators with complicated spectra, the completeness problem is replaced by a problem on the existence of a sufficiently complete collection of invariant subspaces.

Here the operators in the Macaev class play an exceptional role.

We could have restated this theorem for nonselfadjoint perturbations of a selfadjoint operator. However, we gain generality and completeness if we pass from selfadjoint operators to unitary ones.

Let U *an arbitrary unitary operator, let* $V \in \mathfrak{S}_\omega$ *and let the spectrum of* $T = U + V$ *lie on the unit circle. Then to every arc of the unit circle*

$$\delta_{\alpha,\beta} = \{e^{i\theta}\colon \alpha \leqslant \theta \leqslant \beta\}$$

there corresponds an invariant subspace $\mathfrak{L}_{\alpha,\beta}$ *of* T *with the following properties:*

1. $\mathfrak{L}_{\alpha,\beta}$ *is the maximal invariant subspace of* T *for which the spectrum of* T *lies on the arc* $\delta_{\alpha,\beta}$.

2. *If a finite system* $\{\delta\}$ *of open arcs* δ *covers the unit circle then the algebraic sum of the corresponding subspaces* $\{\mathfrak{L}_\delta\}$ *is dense in* $\mathfrak{H}$.

As Macaev [3b] has shown, if an operator T satisfies the hypotheses of the theorem then the following "bilogarithmic condition" is satisfied:

$$\int_{1-\varepsilon}^{1+\varepsilon} \ln^+ \ln^+ M_T(\rho)\, d\rho < \infty \qquad (\varepsilon > 0), \tag{1}$$

where

$$M_T(\rho) = \max_{|\zeta|=\rho} \| (T - \zeta I)^{-1} \| \qquad (\rho \neq 1).$$

According to research by Ju. I. Ljubič and Macaev, condition (1) guarantees the properties of T given by the theorem. The work of these authors implies that the bilogarithmic condition (1) is equivalent to

$$\sum_{n=-\infty}^{\infty} \frac{\ln^+ \| T^n \|}{1+n^2} < \infty. \tag{2}$$

The last condition was first considered in a somewhat more complicated form by J. Wermer.

We must note that the simultaneous parallel work of E. Bishop, the work of C. Foiaş, the preceding work of Wermer and F. Wolf and the recent work of L. de Branges [8a] all closely border on the work of Ljubič and Macaev [7]. The validity of this theorem is proven in the work of Macaev and de Branges. For the statement of this proposition we note that for an invertble operator T the condition $D_T = I - T^* T \in \mathfrak{S}_\omega$ is equivalent to the possibility of representing T in the form $T = U + V$, where U is a unitary operator and $V \in \mathfrak{S}_\omega$.

Let $D \in \mathfrak{S}_\infty \backslash \mathfrak{S}_\omega$. Then under certain restrictions on the dimension of the kernel of D, we can construct an invertible operator T $(\in \mathfrak{R})$ such that $D_T = D$ and in each subspace invariant for T and T^{-1}, the spectrum of T is the whole circle.

§2. Dissipative and contraction operators

Every linearized mechanical system in which there is dissipation of energy is described by a densely defined linear operator A on $\mathfrak{H}$ with values of the form (Af, f) in the left half-plane:

$$\operatorname{Re}(Af, f) \leq 0 \quad (f \in \mathfrak{D}_A).$$

In quantum mechanics, energy dissipation is characterized by the fact that the form of the linear operator describing the physical system lies in the upper half-plane, i.e.

$$\operatorname{Im}(Af, f) \geq 0 \quad (f \in \mathfrak{D}_A).$$

In order to be specific, when speaking of dissipative operators we will have in mind operators of the latter type, i.e. dissipative operators of quantum mechanics.

A dissipative operator is *maximal* if it cannot be extended while preserving its dissipative property. By a theorem of Ralph Phillips, in order for a dissipative operator to be maximal, it is sufficient that it have at least one regular point in the lower half-plane, and necessary that all interior points of this half-plane be regular.

Phillips also showed that every dissipative operator can be extended to a maximal one.

As in the von Neumann theory of the extension of Hermitian operators, the Cayley transform

$$T = (A - iI)(A + iI)^{-1}$$

plays an essential role in Phillips' research. If A is a maximal dissipative operator then (and only then) the Cayley transform gives an operator defined on all of $\mathfrak{H}$ with norm $\|T\| \leq 1$. Such operators are called *contractions*. As a rule, problems about maximal dissipative operators can be restated as problems about contractions.

We will speak mainly of contractions, since the concept of a contraction is more elementary.

A contraction is called simple if it induces a nonunitary operator on each invariant subspace. By a theorem of Heinz Langer, B. Sz.-Nagy and C. Foiaş, every contraction is the orthogonal sum of a unitary operator and a simple contraction.

For our purpose we may limit ourselves to simple contractions and to simple dissipative operators, which are analogously defined.

A contraction T is called *weak* if it satisfies these two conditions:

a) T is invertible: $T^{-1} \in \mathfrak{R}$.

b) The deviation $D_T (= I - T^* T)$ of T from a unitary operator is nuclear.

Condition a) could have been replaced by a more general statement, namely that T have at least one regular point inside the unit disc.

Although condition b) is rather restrictive, nonetheless in passing from weak contractions to the corresponding dissipative operators we obtain a class of operators containing, in particular, the operators generated by Schrödinger's radial equation with complex potential of sufficiently general type.

In research in the theory of general linear operators, the problem of constructing a continuous analog of the theory of elementary divisors has, like the bluebird, remained elusive. It was in searching for this bird that I. M. Gel'fand arrived at the theory of commutative normed rings, one of the most beautiful creations of modern analysis.

These aspirations led N. Dunford, J. Schwartz and their successors to the development of the theory of spectral operators.

However, until just recently it was difficult to give any general class of operators for which some analog of the theory of elementary divisors could be constructed. There have now been some achievements in that direction, and if we speak of weak contractions we can categorically assert that here we have already reached the Cape of Good Hope.

We note first of all that for weak contractions we can construct a certain ersatz secular determinant, namely

$$d_T(\zeta) = \det (T^* (T - \zeta I)(I - \zeta T^*)^{-1}).$$

It is easily shown that the operator whose determinant we are taking is of the form $I + A$, where $A \in \mathfrak{S}_1$. The determinant has the following representation:

$$d_T(\zeta) = \sqrt{\det(T^*T)}\, B_T(\zeta) \exp\left\{ -\int_0^{2\pi} \frac{e^{it}+\zeta}{e^{it}-\zeta}\, d\omega(t) \right\}, \tag{3}$$

where $B_T(\zeta)$ is the Blaschke product

$$B_T(\zeta) = \prod_j \frac{\lambda_j - \zeta}{1 - \bar{\lambda}_j \zeta} \cdot \frac{|\lambda_j|}{\lambda_j},$$

and $\omega(t)$ is a nondecreasing function normalized by the condition $\omega(0) = \omega(+0) = 0$.

Each element of this representation has an operator-spectral interpretation.

In particular, the numbers $\{\lambda_j\}$ are the eigenvalues of the contraction T, and the number of times each one appears is its algebraic multiplicity (i.e. the dimension of the corresponding root subspace).

If we let $\zeta = 0$ in (3), we obtain

$$\sqrt{\det(T^*T)} = \prod |\lambda_j| \exp\left\{ -\int_0^{2\pi} d\omega(t) \right\}.$$

The following group of assertions is valid.

I. *The space* $\mathfrak{H}$ *is uniquely decomposed into the quasi-direct sum of two subspaces* $\mathfrak{E}$ *and* $\mathfrak{H}_0$, *invariant for* T *and* T^{-1}:

$$\mathfrak{H}=\mathfrak{E}\dot{+}\mathfrak{H}_0 \quad (\mathfrak{E}\cap\mathfrak{H}_0=\{0\};\ \overline{\mathfrak{E}+\mathfrak{H}_0}=\mathfrak{H}),$$

where $\mathfrak{E}$ *is the closed linear hull of the root subspaces of* T. *The spectrum of* T *on* $\mathfrak{H}_0$ *lies on the unit circle.*

The system of root vectors of T will be complete, i.e. $\mathfrak{H} = \mathfrak{E}$, if and only if $d_T(\zeta) = B_T(\zeta)$, i.e. $\det(T^*T) = \Pi|\lambda_j|^2$. Thus the following three assertions are equivalent:

$$\mathfrak{H}=\mathfrak{E} \Leftrightarrow d_T(\zeta)=B_T(\zeta) \Leftrightarrow \det(T^*T)=\prod|\lambda_j|^2.$$

We denote the restriction of a contraction T to $\mathfrak{H}_0$ by T_0. Let $\mathfrak{L}_t$ be the maximal invariant subspace of T_0 whose spectrum lies on the arc $e^{i\theta}(0 \le \theta \le t)$. Then

$$\omega(t+0)=-\frac{1}{2}\ln\det(I+P_tD_TP_t), \tag{4}$$

where P_t is the orthogonal projection of $\mathfrak{H}$ onto $\mathfrak{L}_T$ and $D_T = I - T^*T$.

An operator $A \in \mathfrak{R}$ is unicellular if the set of its invariant subspaces is ordered by inclusion. For an operator on an n-dimensional space unicellularity means that the operator consists of one Jordan cell.

If the deviation D_T of a unicellular operator $T(\in \mathfrak{R})$ belongs to the Macaev class, then, as is easily shown from the above, the spectrum of T consists of a single point lying on the unit circle.

The spectrum of a unicellular operator on an n-dimensional space also consists of a single point. But this is not a complete characterization. One complete characterization of such an operator is that its resolvent has a pole of order n, the maximal possible order. It turns out that among the weak contractions, the unicellular operators are also completely characterized by the highest order of growth of the resolvent. More precisely:

II. *For a simple weak contraction* T *to be unicellular it is necessary and sufficient that* 1) T *have no eigenvalues inside the unit disc and* 2) *the equation*

$$\varlimsup_{\rho\uparrow 1}\{(1-\rho)\ln M_T(\rho)\}=-\frac{1}{2}\ln\det(T^*T), \tag{5}$$

be satisfied, where, as above,

$$M_T(\rho)=\max_{|\zeta|=\rho}\|(T-\zeta I)^{-1}\| \qquad (\rho\neq 1).$$

If the first condition is satisfied but T *is not unicellular, then* (5) *holds with a* < sign.

The theorem appears in this form in Gohberg and Kreĭn's second book [6b]. It is a refinement of the Brodskiĭ-Kisilevskiĭ criterion for the unicellularity of a Volterra operator. An operator A is a Volterra operator if it is completely continuous and its spectrum is concentrated at the origin.

Without resorting to the notion of a characteristic function for the moment, we can state the Brodskiĭ-Kisilevskiĭ theorem [9] as follows:

II′. *Let A be a dissipative Volterra operator with nuclear imaginary part, i.e.*

$$A \in \mathfrak{S}_\infty, \quad \sigma(A) = \{0\}, \quad A_J = \frac{1}{2i}(A - A^*) \geqslant 0, \quad \operatorname{sp} A_J < \infty.$$

Then A is unicellular if and only if

$$\lim_{r \downarrow 0} r \ln \|(A - irI)^{-1}\| = 2 \operatorname{sp} A_J.$$

In linear algebra, a linear operator A is unicellular if and only if its spectrum is a single point and it is cyclic, i.e. there is a vector f such that the system $f, Af, \cdots, A^{n-1}f$ is a basis for the space. It is easy to show that every bounded unicellular operator is cyclic, i.e. there exists a vector f such that

$$\bigvee_{n=1}^{\infty} A^n f = \mathfrak{H}.$$

However, in an infinite-dimensional Hilbert space, even if we limit ourselves to Volterra operators, a cyclic operator is not, in general, unicellular. Also, as Kisilevskiĭ has recently shown, every cyclic simple dissipative Volterra operator with nuclear imaginary part is unicellular.

Every linear operator on a finite-dimensional space either is unicellular or decomposes into the direct sum of unicellular operators. Kisilevskiĭ [10] has extended this proposition to dissipative Volterra operators on a Hilbert space.

III. *Every simple dissipative Volterra operator with nuclear imaginary part is either unicellular or*

a) *$\mathfrak{H}$ is the quasi-direct sum of subspaces $\mathfrak{L}_k$ $(k = 1, 2, \cdots, \omega \leq \infty)$, invariant under A:*

$$\mathfrak{H} = \Sigma \underset{\smile}{+} \mathfrak{L}_k, \quad A\mathfrak{L}_k \subseteq \mathfrak{L}_k;$$

and b) *the restriction of A to $\mathfrak{L}_k$ $(k = 1, 2, \cdots)$ is unicellular.*

The cardinal number ω *and the positive numbers*

$$\tau_k = \mathrm{sp}\,(A \mid \mathfrak{L}_k)$$

are invariants of A.

Thus many basic propositions of linear algebra have analogs that are true for simple dissipative Volterra operators with nuclear imaginary part. However, we must especially underline one striking fact, namely that these analogs are true for operators which have no analog in linear algebra. Indeed, if $\sigma(A) = \{0\}$ in linear algebra, then

$$\mathrm{sp}\, A = \mathrm{sp}\, A_R + i\, \mathrm{sp}\, A_J = 0,$$

so $\mathrm{sp}\, A_J = 0$, and since $A_J \geq 0$, we have $A_J = 0$. Thus A is selfadjoint, i.e. there are no simple dissipative Volterra operators on a finite-dimensional space.

Similarly, the notion of a simple contraction with spectrum on the unit circle is absurd in linear algebra.

Also, the most remarkable facts in contraction theory concern simple contractions with spectra on the unit circle.

We have not exhausted the results obtained for unicellular operators. We note that related results for contractions, some equivalent, some supplementary, have recently been obtained by Sz.-Nagy and Foiaș in their remarkable series of articles on contractions.[1)]

In studying the spectral properties of an operator in the algebraic case it is impossible to avoid a certain secular determinant. It is all the more impossible to avoid its surrogate in the infinite-dimensional case. In the preceding assertions we had to turn repeatedly to the resolvent of the operator. It turns out that there is an object, more complicated to determine than the resolvent, that carries the key to the solution of many problems in the theory of nonselfadjoint operators. In particular, a significant part of the preceding assertions was obtained with the aid of this object.

We are talking about the characteristic function of a nonselfadjoint operator. Presently there are various definitions of the characteristic function—of this "sputnik" of a nonselfadjoint operator. We can distinguish two types of characteristic functions. The functions of the first type characterize

1) This series has been published in *Acta Math. Szeged*, beginning in 1953. It consists of 13 articles at present, but we will refer to only some of them ([11], [20]) in the sequel. (See also [26].)

the deviation of the operator from a unitary operator while those of the second type characterize its deviation from a selfadjoint operator. The characteristic function $\theta_T(\zeta)$ of the first type of an operator T is defined by

$$\theta_T(\zeta) = [\,|I - TT^*|^{-1/2}(T - \zeta I)(I - \zeta T^*)^{-1}|I - T^*T|^{1/2}]\,|\,\overline{D_T\mathfrak{H}},$$

where $|A|$ denotes the operator modulus of a selfadjoint operator A, i.e. the nonnegative operator whose square is A^2, and $A|\mathfrak{L}$ denotes the restriction of A to the subspace $\mathfrak{L}$. For each ζ, such that $(I - \zeta T^*)^{-1} \in \mathfrak{R}$, $\theta_T(\zeta)$ is a bounded linear operator mapping $\overline{D_T\mathfrak{H}}$ into $\overline{D_{T^*}\mathfrak{H}}$.

Under rather general conditions the characteristic operator-function gives a simple operator up to unitary equivalence.

This function was first introduced in 1944 in M. S. Livšic's doctoral dissertation for the case

$$\dim D_T\mathfrak{H} = 1.$$

In this case, $\theta_T(\zeta)$ is essentially a scalar function and T itself will be either a dilation or a contraction. In general, when the deviation operator D_T is finite dimensional, $\theta_T(\zeta)$ can be regarded as a matrix-function. For this case, it was essentially obtained in 1950 by Livšic and V. P. Potapov [13]. Unfortunately, lack of time does not permit me to present the evolution of this most important concept, an evolution extended over twenty years under the influence and with the participation of the author of the concept, and which continues today.

In particular, I cannot dwell on the characteristic function of the second type, also first introduced by Livšic, which obtained its essential development in the work of M. S. Brodskiĭ [14, 15a].

In 1963 Sz.-Nagy and Foiaş [11a] arrived at the characteristic function from a quite different situation. They first studied the characteristic function without assuming that the deviation operators D_T and D_{T^*} are completely continuous.

When T is a contraction, the definition of $\theta_T(\zeta)$ is simpler:

$$\theta_T(\zeta) = [T - \zeta D_{T^*}^{1/2}(I - \zeta T^*)^{-1}D_T^{1/2}]\,|\,\overline{D_T\mathfrak{H}}.$$

In this case, the operator function $\theta_T(\zeta)$ is holomorphic inside the unit disc and

$$\|\theta_T(\zeta)\| \leqslant 1 \qquad (|\zeta| \leqslant 1).$$

Multiplying an arbitrary bounded linear operator by a sufficiently small constant, we can turn it into a contraction with arbitrarily small norm. Then we can construct the characteristic function of this contraction. However we can hardly count on the characteristic function's furnishing us with valuable new information. Evidently this function will be useful only when the operator is, in some sense, close to a unitary one. In the work of Soviet mathematicians, the measure of such closeness was, until recently, the condition that the deviation operators D_T and D_{T^*} be completely continuous. But there is a closeness of another sort, namely that T be similar to a unitary operator, meaning the existence of a bounded invertible operator $S\,(S^{-1} \in \Re)$ such that

$$T = S^{-1}US, \tag{6}$$

where U is a unitary operator. It was discovered comparatively recently that a simple contraction can be similar to a unitary operator. In the algebraic case every contraction similar to a unitary operator is itself unitary. The great achievement of the Sz.-Nagy –Foiaş theory is the following general criterion for a contraction to be similar to a unitary operator.

IV. *A contraction T is similar to a unitary operator if and only if for every $\zeta(|\zeta| < 1)$ the operator $\theta_T(\zeta)$ is a one-to-one mapping of $\overline{D_T\mathfrak{H}}$ onto $\overline{D_{T^*}\mathfrak{H}}$ and*

$$\sup_{|\zeta|<1} \|\theta_T^{-1}(\zeta)\| < \infty.$$

It is easy to see that if some operator T (not necessarily a contraction) is similar to a unitary operator, then it possesses the following two properties:

$$\alpha)\qquad \sigma(T) \subset \{e^{i\varphi}\colon 0 \leqslant \varphi \leqslant 2\pi\}$$

and

$$\beta)\qquad \sup_{|\zeta|\neq 1} (1-|\zeta|)\,\|(T-\zeta I)^{-1}\| < \infty.$$

As a corollary of the Sz.-Nagy –Foiaş criterion, when T is a contraction these conditions are not only necessary but also sufficient for T to be similar to a unitary operator [6a].

As A. S. Markus has noted, we can construct an operator having a previously specified spectrum on the unit circle that satisfies α) and β) but that is not similar to a unitary operator. It is understood that such an operator will not be a contraction.

The Sz.-Nagy–Foiaş criterion for weak contractions gives the following sufficiency test:

A weak contraction T with spectrum on the unit circle is similar to a unitary operator whenever its spectral shift function $\omega(t)$ belongs to the class $\mathrm{Lip}_1(0, 2\pi)$.

If the deviation operators are finite dimensional then this test is again sufficient. We can indicate still another case when this test is sufficient. To this end we require a precise analysis of the structure of the invariant subspaces of the operator.

If a contraction T is similar to a unitary operator then

$$T = \int_0^{2\pi} e^{it}\, dF(t),$$

where $F(t)$ is some skew resolution of the identity, normalized, for example by $F(t-0) = F(t)$ $(0 < t \leq 2\pi)$, $F(0) = 0$, $F(2\pi) = I$. Let $P(t)$ denote the orthogonal projection of $\mathfrak{H}$ onto $F(t)\mathfrak{H}$ $(0 \leq t \leq 2\pi)$. The resolution of the identity $P(t)$ $(0 \leq t \leq 2\pi)$ is called the rectified spectral function T. We have the following theorem.

Let H be a nonnegative operator, $\|H\| < 1$, and let $P(t)$ $(0 \leq t \leq 2\pi)$ be an orthogonal resolution of the identity.

There exists a contraction T that is similar to a unitary operator, has deviation $D_T = H$ and rectified spectral function $P(t)$ $(0 \leq t \leq 2\pi)$ if and only if the operator-function $H^{1/2} P(t) H^{1/2}$ satisfies the Lipschitz condition

$$\| H^{1/2} (P(t) - P(s)) H^{1/2} \| \leqslant \mathrm{const}\, | t - s |.$$

When this condition is satisfied, the operator $T = U(I - V)^{-1}$, where

$$U = \int_0^{2\pi} e^{it}\, dP(t)$$

and

$$V = \int_0^{2\pi} (I - P(t) H P(t))^{-1} P(t) H\, dP(t), \tag{7}$$

is the unique contraction with the properties:

1) *the spectrum of T lies on the unit circle;*

2) *$P(t+0)\mathfrak{H}$ $(0 < t < 2\pi)$ is the maximal invariant subspace of T for*

which the spectrum lies on the arc $e^{i\phi}\,(0 \le \phi \le t)$.

We omit the statement of an analogous theorem describing dissipative operators that are similar to selfadjoint ones.

As an example, consider the following contraction on the space of n-dimensional vector-functions $L_2^{(n)}(0, 1)$:

$$(Tf)(t) = e^{i\alpha(t)} f(t) + \int_t^1 K(t, s) f(s)\, ds,$$

where $K(t, s)$ $(0 \le t, s \le 1)$ is (for simplicity) a continuous Hermitian matrix kernel inducing a nuclear operator. Then $\omega(t)$ is computed from the formula

$$\omega(t) = \int\limits_{0 \le \alpha(s) \le t} \operatorname{sp} K(s, s)\, ds.$$

That $\omega(t)$ belong to Lip_1 is a necessary and sufficient condition for T to be similar to a unitary operator.

For the dissipative operator

$$(Af)(t) = \alpha(t) f(t) + 2i \int_t^1 K(t, s) f(s)\, ds \quad (f \in L_2(0, 1)),$$

where $K(t, s)$ is an arbitrary continuous Hermitian positive kernel, this condition ($\omega \in \mathrm{Lip}_1$) is necessary and sufficient for A to be similar to a selfadjoint operator.

For $n = 1$ and $K(t, s) \equiv 1$, this last result was obtained earlier in [11c]. When $\alpha(t) = t$, A will always be similar to a selfadjoint operator [16b].

§3. Triangular representation of operators and multiplicative representation of operator-functions

The above theorem, giving a complete description of those contractions that are similar to unitary operators, was obtained by Gohberg and Kreĭn [6d] using both the results of Sz.-Nagy and Foiaş that we have mentioned and new methods involving the theory of abstract triangular and multiplicative integrals. A triangular integral (the right side of (7)) appears in the very statement of the theorem.

We now give a series of definitions. A closed (in the sense of strong convergence) set of orthogonal projections $\mathfrak{P} = \{P\}$ is called a *chain* if it is ordered (in the natural way) and contains the projections 0 and I. A pair of projections (P^-, P^+) $(P^- < P^+;\ P^\pm \in \mathfrak{P})$ is a *gap* in the chain $\mathfrak{P}$ if $\mathfrak{P}$ contains no projection lying between P^- and P^+. A chain without gaps is said to be *continuous*.

A chain $\mathfrak{P}$ is *maximal* if it is not a proper subset of any other chain. A chain $\mathfrak{P}$ is a *proper* chain of a given operator $A\,(\in \mathfrak{R})$ if each subspace $P\mathfrak{H}$ is invariant with respect to A, i.e. if $AP = PAP\ (P \in \mathfrak{P})$.

By a theorem of Hilbert, every selfadjoint bounded linear operator H admits the representation

$$H = \int_a^b \lambda\, dE_\lambda,$$

where E_λ is the resolution of the identity induced by H. Let $\mathfrak{P}$ be an arbitrary proper chain of H containing the chain $\{E_\lambda\}_{a \le \lambda \le b}$. Then the last integral can be rewritten in the form

$$H = \int_{\mathfrak{P}} \lambda(P)\, dP = \int_{\mathfrak{P}} \lambda(P)\, P\, dP,$$

where $\lambda(P)$ is the upper bound of the spectrum of H on the invariant subspace $P\mathfrak{H}$. This integral could have been called the diagonal integral along the chain. It is a special case of a triangular integral along the chain, which has the form

$$J = \int_{\mathfrak{P}} F(P)\, dP, \tag{8}$$

where $F(P)$ is an operator-valued function on the chain.

The integral (8) is understood in the sense of convergence of the corresponding partial sums in the operator norm, and the limit is taken along the directed set of all partitions of $\mathfrak{P}$.

If

$$F(P) = PF(P),$$

then (8) is called a triangular integral. In this case $\mathfrak{P}$ will be a proper chain for the operator J.

In the finite-dimensional case, a triangular integral in a basis that expands together with the chain $\mathfrak{P}$ will be represented by a triangular matrix.

The simplest triangular integral, naturally, is

$$\int_{\mathfrak{P}} PC\, dP, \tag{9}$$

where C is a completely continuous operator. This integral first appeared in 1958 in an article by Brodskiĭ, as a result of a revision of Livšic's triangular model of a nonselfadjoint operator. Brodskiĭ showed that every Volterra operator A has the representation

$$A = \int_{\mathfrak{P}} PC\, dP,$$

where $\mathfrak{P}$ is some proper maximal chain of A and $C = 2iA_j$ (the existence of such a chain follows from the results of von Neumann and Aronszajn and Smith [17]).

However, the following question was not solved by this representation: let an operator C and a chain $\mathfrak{P}$ be given; under what conditions does the integral (9) exist, and, if it exists and if C belongs to such-and-such a class of operators, then to what class does the integral belong? In connection with this question certain rather unexpected relations among the spectra of the Hermitian parts of Volterra operators were discovered. The Macaev class turned out to play an exceptional role here. The integral (9) converges along any continuous chain $\mathfrak{P}$ if and only if $C \in \mathfrak{S}_\omega$. The most difficult part of this theorem, the sufficiency part, was proved by Macaev.

The theory of the triangular integral (9), together with its application to selfadjoint boundary value problems for canonical systems of differential equations, can be found in the second Gohberg-Kreĭn book [6b]. Applications of the theory of this integral to stability problems for the solutions of canonical systems of differential equations with periodic Hamiltonian can also be found in this book.

We cannot dwell on more general triangular representations of operators—representations still containing the diagonal part of an operator.

At the base of the theory of the triangular integral lies the notion that there are certain continuous analogs of I. Schur's elementary theorem on the possibility of reducing any matrix to triangular form with a unitary transformation.

An entire century before Schur, Lagrange gave a method of reducing a quadratic form to a sum of squares. In matrix theory language, this method gave, in particular, a solution to the problem of representing a positive definite matrix in the form

$$G = A^* A, \tag{10}$$

where A is a triangular matrix. The continuous analog of this problem in general form was considered by Gohberg and Kreĭn [6].

In the operator case, the problem is stated as follows: for simplicity, let G be a positive definite operator and $\mathfrak{P}$ a chain; we are required to represent G in the form (10), where $A\,(\in \mathfrak{R})$ is some operator with proper chain $\mathfrak{P}$: $PAP = AP\,(P \in \mathfrak{P})$. We will call this the problem of factoring G along $\mathfrak{P}$.

If this problem has a solution then A is obviously determined up to a right unitary multiplier commuting with the projections $P \in \mathfrak{P}$. However if $G = I + H$, where $H \in \mathfrak{S}_\infty$, and if a multiplier A is found in the same form $A = I + X$, $X \in \mathfrak{S}_\infty$, then, in the factorization (10), if A exists it is uniquely determined.

The operator A is expressed in terms of H and $\mathfrak{P}$ by the triangular integral

$$(A-I)^{-1} = I + \int_{\mathfrak{P}} (I - PHP)^{-1} PH\, dP \tag{11}$$

or by the multiplicative[1] integral

$$A = \overset{\curvearrowleft}{\int_{\mathfrak{P}}} \exp\left(dPHP\,(I - PHP)^{-1}\right).$$

---

1) We note in general that under broad conditions the triangular integral

$$J = \int_{\mathfrak{P}} F(P)\, dP$$

is related to the multiplicative integral by the following equality:

$$(J+I)^{-1} = \overset{\curvearrowleft}{\int_{\mathfrak{P}}} \exp\left(dPF(P)\right).$$

The last integral is understood as the limit in norm of the corresponding partial products along the directed set of partitions of the chain.

It turns out that if H belongs to the Macaev class then the integrals in question converge along any continuous chain $\mathfrak{P}$. Although Gohberg and Kreĭn succeeded in showing that $C \in \mathfrak{S}_\omega$ is also a necessary condition in the case of the simpler triangular integral (9), the analogous question here remains open.

We should emphasize that for a fixed continuous chain $\mathfrak{P}$ the integral (9) (or (11)) may converge even when C (H, respectively) is not completely continuous. This will happen, for example, if $\mathfrak{P}$ is sufficiently smooth with respect to $C\,(H)$. In this case we encounter (7).

Without going into details, we point out that the factorization in the theory of Wiener-Hopf-type equations fits into the general scheme that we have mentioned of the abstract problem of factorization along a chain.

In the research of Gohberg and Kreĭn the factorization problem was not an end in itself. It arose as a certain intermediate step, necessary for the construction of a triangular representation theory for operators close to unitary ones. The solution of this problem played an important role in obtaining a multiplicative representation of the characteristic operator-function of an operator close to a unitary one. For simplicity, we will limit ourselves to the case of a contraction T with no part of its spectrum inside the unit disc. Such a representation may be written in the following form:

$$T^*\theta_T(\zeta) = |\,T\,| \overset{\curvearrowleft}{\int_{\mathfrak{P}}} \exp\left(-\frac{e^{i\varphi(P)}-\zeta}{e^{i\varphi(P)}+\zeta}\right) \mathfrak{R}^*(P)\, dP\, \mathfrak{R}(P), \tag{12}$$

where $\mathfrak{P}$ is a proper maximal chain of T containing the chain P_T, which has already been determined, and $\mathfrak{R}(P)$ is some operator function on $\mathfrak{P}$, computed according to completely determined rules. In the case of a weak contraction or of a contraction similar to a unitary operator, the integral is determined by the usual means. In other cases, for example, when $D_T \in \mathfrak{S}_\omega$ (and in this case (12) always holds), the integral must in general be determined by special means.

We note that V. M. Brodskiĭ (the younger) took part in this research of Gohberg and Kreĭn. It was influenced by the earlier research of M. S. Brodskiĭ [15a,b], who first succeeded in obtaining a multiplicative representation of the characteristic function (of the second type) for a bounded operator with completely continuous imaginary part as a corollary of the triangular representation.

We note that since the proper chain $\mathfrak{P}_T$ is in general not uniquely determined for an operator, the representation (12) of its characteristic function is not uniquely determined.

The analytic problems concerning the description and classification of all multiplicative representations of the characteristic operator-function of a given operator evidently must constitute a theory which replaces the theory of elementary divisors in passing from the finite-dimensional to the infinite-dimensional case.

This purely operator research was preceded by the fundamental research of V. P. Potapov [18] and his student Ju. P. Ginzburg [19] into the multiplicative representation of analytic matrix- and operator-functions. We will state Potapov's theorem for the so-called definite case.

Let $\theta(\zeta)$ be a contraction-valued matrix-function that is holomorphic in the unit disc. Then $\theta(\zeta)$ admits the following representation:

$$\theta(\zeta) = U \overset{\curvearrowleft}{\int_a^b} \exp\left\{-\frac{e^{i\varphi(t)}+\zeta}{e^{i\varphi(t)}-\zeta}\, dF(t)\right\} B(\zeta), \tag{13}$$

where

$$B(\zeta) = \overset{\curvearrowleft}{\prod_k} \left(\frac{\zeta_k-\zeta}{1-\bar{\zeta}_k\zeta}\cdot\frac{|\zeta_k|}{\zeta_k} P_k + Q_k\right),$$

U is a fixed unitary operator, $\phi(t)$ is a nondecreasing function $(0 \le \phi(t) \le 2\pi$ for $a \le t \le b)$, $F(t)$ is a monotone Hermitian-valued matrix-function, the P_k are orthogonal projections and $Q_k = I - P_k$.

This theorem generalizes the corresponding theorem on the multiplicative representation of a bounded scalar function holomorphic in the disc. We essentially used this last theorem in writing out the multiplicative representation of the determinant $d_T(\zeta)$ for a weak contraction. A wider class of scalar functions is the class introduced by R. Nevanlinna, the so-called class of functions of bounded type. The classes of F. Riesz, M. Riesz, Hardy and others are subclasses of this class. Ginzburg succeeded in determining matrix and operator analogs of these classes and obtained multiplicative representations for them.

How are the theorems of Potapov and Ginzburg related to the theorems on the multiplicative representation of the characteristic operator-function? They are very closely related, although at the present time the results do not overlap. The fact is that every function $\theta(\zeta)$, holomorphic in the disc, whose values are contractions on some space $\mathfrak{N}$, admits the following representation:

$$\theta(\zeta) = V\theta_T(\zeta) \oplus U, \qquad (14)$$

where V and U are fixed unitary operators and $\theta_T(\zeta)$ is the characteristic function of some simple contraction T. T is determined to within unitary equivalence and acts on some new space $\mathfrak{H}$.

The representation (14), even in the case of a scalar function $\theta(\zeta)$, where $U = 0$ and V is a number of modulus one, constituted something of a discovery, due to Livšic in his above-mentioned 1944 work [12a]. Afterwards, it was generalized by that author, his students and colleagues. Finally this representation was established in its most general form by Sz.-Nagy and Foiaş. A series of results and the general idea of a connection between the multiplicative representation of a matrix-function or operator-function and the structure of the invariant subspaces of some nonselfadjoint operator also belong to Livšic.

The theorems of Ginzburg on the uniqueness of the multiplicative representation (13) of contractible matrix-functions were a great success, and he extended these theorems to the so-called nuclear case. It is characteristic that to obtain a definite part of these results it was necessary to use the connection with the theory of the characteristic function of an operator and the theorems on the unicellularity of weak contractions. Finally, we note that Potapov also obtained a multiplicative representation for matrix-functions that are meromorphic in the unit disc and whose values are contraction operators with respect to a certain indefinite metric. These results, and especially the following generalizations of Ginzburg, have so far not been completely obtained by purely operator methods.

We must hope that in the near future the divergence between the purely analytic methods in the multiplicative representation problem and, on the other hand, the operator methods based on the triangular representation and the characteristic function theory, will be liquidated.

To a significant extent, this divergence is explained by the fact that for operators that are close to unitary ones and are not contractions or dilations, there is evidently still no sufficiently general definition of the characteristic operator-function. The fact is that for the present definitions of this function, there is no decomposition analogous to (14) for the indefinite case.

It is understood that when we speak of the multiplicative representation of operator-functions, the division of methods into purely analytic and purely operator methods is largely conventional. In fact, the object in question is a hybrid: it is an analytic function, but its values are operators.

In particular, this hybrid has an especially complicated operator character in the indefinite case. We are unable to dwell on the difficulties generated by the indefiniteness. They are very unpleasent even in the finite-dimensional case.

The Sohockiĭ-Plemelj formulas play an important role in the theory of the Stieltjes-Cauchy integral. L. A. Sahnovič [16a, b] obtained analogs of these formulas for multiplicative integrals under rather broad assumptions. Under sufficiently general assumptions on the character of the dissipative perturbations, these formulas allowed him to construct a scattering theory, very similar to the Rosenblum-Kato scattering theory for selfadjoint operators. I would also like to mention the recent work of Tosio Kato on the theory of Schrödinger equations with complex potential [2], which overlaps these results. So far I have not mentioned that the starting point for the Sz.-Nagy–Foiaş investigations is Sz.-Nagy's theorem on the existence of a unitary dilation for every contraction. The successive development of the dilation theory of contractions shows its relation to scattering problems and to the prediction theory of stationary random processes.

As a result it was clarified that the characteristic function of a contraction can always be treated as the Heisenberg S-matrix of a suitable scattering problem.

This was all discovered when Lax and Phillips constructed a new scattering theory of acoustic waves on obstructions. This elegant theory is remarkable for its bringing together the methods of classical mathematical physics, abstract operator theory and ideas taken from quantum mechanics and the prediction theory of stationary random processes. In this theory, a new abstract scheme for scattering problems was proposed which immediately brought the work of these authors and that of Sz.-Nagy and Foiaş closer together. The relation between the two theories is clearly explained in recent papers by two young mathematicians, V. M. Adamjan and D. Z. Arov [25a, b, c, d]. In these papers part of the work of the four authors of whom we have spoken is generalized (in the direction of continuing to bring it closer to the prediction research) in the form of the distinctive theory of the chaining of semi-unitary operators.

In the very near future we will have the satisfaction of knowing the

Sz.-Nagy–Foiaş and Phillips-Lax theories in detail from books by these authors [21], [26].

Here we also mention Livšic's recent book [12c] which gives applications of the characteristic matrix-function to various problems in electric network theory, wave guides and nuclear physics.

§4. Entire Hermitian operators and the theory of canonical representations of operators

The preceding discussions had an especially nonselfadjoint character. Also, part of the above is of cardinal importance in that research in the theory of selfadjoint operators which comes under the title "theory of canonical representations of selfadjoint operators".

At the end of the thirties, after the work of T. Carleman, von Neumann, Stone and M. A. Naĭmark, the theory of Hermitian and selfadjoint operators seemed complete; the impression was given that only the penetration of the completed theory into the adjacent domains of mathematatices and physics remained for succeeding researchers and generations. In fact, difficult problems lay ahead, requiring a new and greater analytic apparatus.

In order to give the first representation of one of the problems just mentioned, we recall the following elementary fact from linear algebra. As is well known, for every Hermitian operator on a finite-dimensional space with simple spectrum there exist an infinite number of orthonormal bases in which the operator is represented by a Jacobi (i.e. tridiagonal) Hermitian matrix.

This result immediately carries over to bounded operators. Let $G\,(\in \mathfrak{R})$ be a selfadjoint operator with simple spectrum. The simplicity of the spectrum means that G is a cyclic operator, i.e. there is a vector $u \in \mathfrak{H}$ for which the sequence $u, Gu, G^2u, \cdots$ is complete in $\mathfrak{H}$. If we orthogonalize this sequence by the Schmidt process, then in the basis obtained G will correspond to a Jacobi matrix

$$G \sim \left\| \begin{matrix} a_1 & b_1 & 0 & & \\ \bar{b}_1 & a_2 & b_2 & 0 & \\ 0 & \bar{b}_2 & a_3 & \ddots & \\ & & \ddots & \ddots & \ddots \end{matrix} \right\| .$$

This representation is not unique and is determined by the choice of the generating vector u.

What happens if our operator $G = H$ is unbounded? It is trivial that an orthonormal basis in which the operator is represented by a Jacobi matrix does not exist for every operator with simple spectrum.

On the other hand, it became known recently that second order differential operators must be considered as continuous analogs of operators given by Jacobi matrices. Is the supply of differential operators sufficient to obtain a representation of an arbitrary selfadjoint operator H for any previously given generating element u? The supply, in general, turns out to be insufficient, even if we admit generalized functions as the coefficients of a differential operator. We obtain the required class if we introduce the canonical differential operators. With the help of these differential operators we can describe all selfadjoint operators, with spectrum of any multiplicity. The notion of an entire Hermitian operator is central in this theory.

We give the definition of an entire operator, more precisely an $\mathfrak{L}$-entire Hermitian operator. Let H be a simple closed Hermitian operator with dense domain of definition $\mathfrak{D}(H)$. Let $\mathfrak{M}_z$ denote the set of values of the operator $H - zI$:

$$\mathfrak{M}_z = (H - zI)\,\mathfrak{D}(H).$$

Let $\mathfrak{L}$ be a subspace of $\mathfrak{H}$. The given simple Hermitian operator H is called $\mathfrak{L}$-entire if for every z

$$1)\ \mathfrak{M}_z = \overline{\mathfrak{M}}_z \ \text{ and } \ 2)\ \mathfrak{M}_z \dotplus \mathfrak{L} = \mathfrak{H}.$$

We note that just the first condition implies that the defect numbers of H are the same, i.e. $\dim(\mathfrak{H} \ominus \mathfrak{M}_z) = \dim \mathfrak{L}$. Let $\mathfrak{E}(z)$ denote the skew projection of $\mathfrak{H}$ onto $\mathfrak{L}$ parallel to $\mathfrak{M}_z$. If H is $\mathfrak{L}$-entire then $\mathfrak{E}(z)$ turns out to be an entire operator-function. This means that the mapping $f \longrightarrow f_{\mathfrak{L}}(z) = \mathfrak{E}(z)f$ associates with each vector $f \in \mathfrak{H}$ an entire function of z with values in $\mathfrak{L}$. This mapping is one-to-one, and it carries H into the operator of multiplication by z, i.e. if $g = Hf$ then $g_{\mathfrak{L}}(z) = zf_{\mathfrak{L}}(z)$. This explains the appellation "$\mathfrak{L}$-entire operator".

In this representation the scalar product will be written in the following form:

$$(f,\ g) = \int_{-\infty}^{\infty} g_{\mathfrak{L}}^{*}(\lambda)\, d\sigma(\lambda)\, f_{\mathfrak{L}}(\lambda) \qquad (f,\ g \in \mathfrak{H}), \tag{15}$$

where $\sigma(\lambda) = \sigma(\lambda - 0)$, with $\sigma(-\infty) = 0$, is a nondecreasing bounded operator-function whose values are nonnegative operators on $\mathfrak{L}$. If $\mathfrak{L}$ is one dimensional then all these functions may be considered as scalars. To give the right-hand side a more elegant form the following notation for the scalar product of vectors in $\mathfrak{L}$ is used:

$$(a, b) = b^*a \qquad (a, b \in \mathfrak{L}).$$

The set $\Sigma(H; \mathfrak{L})$ of all spectral functions $\sigma(\lambda)$ giving representation (15) is described by the formula

$$\sigma(\lambda) = P_{\mathfrak{L}} E(\lambda) P_{\mathfrak{L}},$$

where $E(\lambda)$ is some generalized spectral function of H and $P_{\mathfrak{L}}$ is the orthogonal projection onto $\mathfrak{L}$. There is a purely analytic description of $\Sigma(H; \mathfrak{L})$, obtained with the help of the so-called second-order resolvent matrix $W(z)$ whose elements are operators on $\mathfrak{L}$. The following compact representation of this matrix was obtained by the author and Š. N. Saakjan in a recent paper [23]:

$$W(z) = \begin{pmatrix} I - zF(z)\mathscr{E}^*(0) & -zF(z)F^*(0) \\ z\mathscr{E}(z)\mathscr{E}^*(0) & I + z\mathscr{E}(z)F^*(0) \end{pmatrix}, \tag{16}$$

where $F(z)$ is the entire operator-function associated with $\mathfrak{E}(z)$; it is defined by

$$F(z) = P_{\mathfrak{L}}(H - zI)^{-1}(1 - \mathscr{E}(z)).$$

The general form of the integral

$$\int_{-\infty}^{\infty} \frac{d\sigma(\lambda)}{\lambda - z} \qquad \left(\sigma(\lambda) \in \Sigma(H; \mathfrak{L})\right).$$

is found with the help of the matrix-function $W(z)$. This result should be considered as a generalization of a classical result of Rolf Nevanlinna which gives a description of all solutions to an indeterminate moment problem. To some extent this is not unexpected. We may consider the spectral function $\sigma(\lambda)$ as a solution to a generalized moment problem. Indeed, the left-hand side of (15) is a known quantity for any f and $g \in \mathfrak{H}$, and the functions $f_{\mathfrak{L}}(z)$ and $g_{\mathfrak{L}}(z)$ are also known. This result touches both on old research by Naĭmark and the author on a description of generalized resolvents, and on succeeding work of A. V. Štraus and his students.

It was unexpected that under rather general conditions $W(1/z)$ is the characteristic function (in the sense of M. S. Brodskiĭ) of some Volterra operator. Thus by the results of Potapov and Brodskiĭ this function admits the multiplicative representation

$$W(z) = \overset{\curvearrowleft}{\int_0^l} \exp(-zJ\mathscr{H}(t)\,dt),$$

where $\mathfrak{H}(t)$ is a function whose values are bounded Hermitian operators on the orthogonal sum $\mathfrak{L} \oplus \mathfrak{L}$ of two copies of $\mathfrak{L}$, and J is the real square root of $-I$:

$$J = \begin{pmatrix} 0 & I_{\mathfrak{L}} \\ -I_{\mathfrak{L}} & 0 \end{pmatrix}.$$

The canonical differential equation

$$J\frac{d\varphi}{dt} = \lambda \mathscr{H}(t)\,\varphi \qquad (0 \leqslant t \leqslant l)$$

corresponds to this represenation. It turns out that the set $\Sigma(H; \mathfrak{L})$ of spectral functions $\sigma(\lambda)$ coincides with the set of spectral functions of an incomplete boundary value problem for this canonical equation. Incompleteness means that the boundary condition is given only at the left end point. The functions $f_{\mathfrak{L}}(z)$ themselves may be interpreted as the generalized Fourier transforms of the corresponding functions in $L_2(0, l; \mathfrak{L})$ using the specific form of the normalized fundamental function $\phi(t; \lambda)$.

Now let H be an arbitrary selfadjoint operator, let

$$H = \int_{-\infty}^{\infty} \lambda\, dE(\lambda)$$

be its spectral resolution and let $\mathfrak{L}$ be some minimal reproducing subspace, i.e.

$$\bigvee_{-\infty<\lambda<\infty} \{E(\lambda)\,\mathfrak{L}\} = \mathfrak{H}.$$

It turns out that H can be considered as the limit of an expanding system of $\mathfrak{L}$-entire operators. In other words, this means that we can associate with H a canonical equation of type (15), however now on the half-line, which has $\sigma(\lambda) = P_{\mathfrak{L}}\, E(\lambda)\, P_{\mathfrak{L}}$ as its unique spectral function.

We see that we succeeded in handling first order equations, but the phase dimension of these equations is equal to twice the dimension of $\mathfrak{L}$ (if $\mathfrak{L}$ is finite dimensional). If H is positive and has simple spectrum then the corresponding canonical equation may be transformed into a string equation with arbitrarily distributed mass. Unfortunately, the Hamiltonian $\mathcal{H}(t)$ is in general not uniquely determined. Obviously, the question of what supplementary conditions must be placed on $\mathcal{H}(t)$ so that it will be uniquely determined is one of great difficulty and we can hardly hope that it will be solved in the next few years.

For the case $n = \dim \mathfrak{L} = 1$, when the spectral function $\sigma(\lambda)$ is a scalar nondecreasing function, we can always get a corresponding canonical system with real Hamiltonian $\mathcal{H}(t)$ (with trace = 1). Nearly fifteen years ago I arrived at the conjecture that with such normalization in this case the Hamiltonian will be uniquely determined. This was verified by me in several important cases. A complete proof of this proposition was recently obtained by Louis de Branges [8b] in his research on Hilbert spaces of entire functions. Likewise my second proposition [24] that *every real simple Volterra operator with two-dimensional skew-symmetric part is a real unicellular operator* was also proven.

I cannot state the various propositions and problems arising in the theory of canonical representations of selfadjoint operators. For the development of this theory it is rather essential to expand it to the case where the space of generalized elements of first-order singularity with respect to H is taken as $\mathfrak{L}$. When $\mathfrak{L}$ is finite dimensional the corresponding generalization is obtained without difficulty. When $\mathfrak{L}$ is infinite dimensional a precise statement of the problem has not yet been worked out. Probably this is connected with the fact that up to now no entire Hermitian operators represented as partial differential operators have been constructed.

Completing my report, I would like to stress that I have fulfilled the responsibility not of an architect but rather of a guide, pointing out the various remarkable sights of a great new center. Sometimes we went slowly, sometimes we hurried, flying headlong through whole buildings. Naturally, with only a short time at my disposal I could not help but pain the leaders of whole regions and territories which we rushed past or even went around. I give them my apologies.[1)]

---

1) I express my deep thanks to I. C. Gohberg for the invaluable aid which he rendered in preparing this report.

BIBLIOGRAPHY

[1] K. O. Friedrichs, *Perturbation of spectra in Hilbert space,* Lectures in Appl. Math., vol. 3, Amer. Math. Soc., Providence, R.I., 1965. MR 32 #365.

[2] T. Kato, *Perturbation theory for linear operators,* Die Grundlehren der math. Wissenschaften, Band 132, Springer-Verlag, New York, 1966. MR 34 #3324.

[3] V. I. Macaev, a) *Function-theoretic methods for certain problems in the theory of selfadjoint linear operators,* Doctoral Dissertation, Moscow, 1966. (Russian)
b) *A class of completely continuous operators,* Dokl. Akad. Nauk SSSR 139 (1961), 548–551 = Soviet Math. Dokl. 2 (1961), 972–976. MR 24 #A1617.

[4] M. V. Keldyš, *On the characteristic values and characteristic functions of certain classes of nonselfadjoint equations,* Dokl. Akad. Nauk SSSR 77 (1951), 11–14. (Russian) MR 12, 835.

[5] M. V. Keldyš and V. B. Lidskiĭ, *On the spectral theory of nonselfadjoint operators,* Proc. Fourth All-Union Math. Congress (Leningrad, 1961), vol. I, Izdat. Akad. Nauk SSSR, Leningrad, 1963, pp. 101–120. (Russian) MR 30 #1414.

[6] I. C. Gohberg and M. G. Kreĭn, a) *Introduction to the theory of linear nonselfadjoint operators in Hilbert space,* "Nauka", Moscow, 1965; English transl., Transl. Math. Monographs, vol. 18, Amer. Math. Soc., Providence, R. I., 1969. MR 36 #3137.
b) *The theory and applications of Volterra operators in Hilbert space,* "Nauka", Moscow, 1967; English transl., Transl. Math. Monographs, vol. 24, Amer. Math. Soc., Providence, R. I., 1970. MR 36 #2007.
c) *On the factorization of operators in Hilbert space,* Acta Sci. Math. Szeged 25 (1964), 90–123; English transl., Amer. Math. Soc. Transl. (2) 51 (1966), 155–188. MR 29 #6313.
d) *On a description of contraction operators similar to unitary operators,* Funkcional. Anal. i Priložen. 1 (1967), no. 1, 38–60 = Functional Anal. Appl. 1 (1967), 33–52.

[7] Ju. I. Ljubič and V. I. Macaev, *On operators with separable spectrum,* Mat. Sb. 56 (98) (1962), 433–468; English transl., Amer. Math. Soc. Transl. (2) 47 (1965), 89–129. MR 25 #2450.

[8] L. de Branges, a) *Some Hilbert spaces of analytic functions.* II, III, J. Math. Anal. Appl. 11 (1965), 44–72; ibid., 12 (1965), 149–186. MR 35 #778, 779.
b) *Some Hilbert spaces of entire functions.* I–IV, Proc. Amer. Math. Soc. 10 (1959), 840–846; Trans. Amer. Math. Soc. 99 (1961), 118–152; 100 (1961), 73–115; 105 (1962), 43–83. MR 22 #4833; 24 #A3289b, A3289c; 26 #583.

[9] M. S. Brodskiĭ and G. È. Kisilevskiĭ, *Criterion for unicellularity of dissipative Volterra operators with nuclear imaginary components,* Izv. Akad. Nauk SSSR Ser. Mat. 30 (1966), 1213–1228; English transl., Amer. Math. Soc. Transl. (2) 65 (1967), 282–296. MR 34 #3310.

[10] G. È. Kisilevskiĭ, *Invariant subspaces of dissipative Volterra operators with nuclear imaginary components,* Izv. Akad. Nauk SSSR Ser. Mat. 32 (1968), 3–23 = Math. USSR Izv. 2 (1968), 1–20. MR 36 #4375.

[11] B. Sz.-Nagy and C. Foiaş, a) *Sur les contractions de l'espace de Hilbert.* VIII: *Fonctions caractéristiques. Modèles fonctionnels,* Acta Sci. Math. Szeged 25 (1964), 38–71. MR 30 #2348.
b) *Sur les contractions de l'espace de Hilbert.* IX: *Factorisations de la fonction caractéristique. Sous-espaces invariants,* Acta Sci. Math. Szeged 25 (1964), 283–316. MR 31 #5101.
c) *Sur les contractions de l'espace de Hilbert.* X: *Contractions similaires à des transformations unitaires,* Acta Sci. Math. Szeged 26 (1965), 79–91. MR 34 #1856.
d) *Sur les contractions de l'espace de Hilbert.* XI: *Transformations unicellulaires,* Acta Sci. Math. Szeged 26 (1965), 301–324. MR 32 #8176.

[12] M. S. Livšic, a) *On a certain class of operators in a Hilbert space,* Mat. Sb. 19 (61) (1946), 239–262; English transl., Amer. Math. Soc. Transl. (2) 13 (1960), 61–83. MR 8, 588; 22 #3981a.
b) *On spectral resolution of linear nonselfadjoint operators,* Mat. Sb. 34 (76) (1954), 145–199; English transl., Amer. Math. Soc. Transl. (2) 5 (1957), 67–114. MR 16, 48; 18, 748.
c) *Operators, oscillations, waves. Open systems,* "Nauka", Moscow, 1966. (Russian) MR 38 #1922.

[13] M. S. Livšic and V. P. Potapov, *A theorem on the multiplication of characteristic matrix functions,* Dokl. Akad. Nauk SSSR 72 (1950), 625–628. (Russian) MR 11, 669.

[14] M. S. Brodskiĭ and M. S. Livšic, *Spectral analysis of non-selfadjoint operators and intermediate systems,* Uspehi Mat. Nauk 13 (1958), no. 1 (79), 3–85; English transl., Amer. Math. Soc. Transl. (2) 13 (1960), 265–346. MR 20 #7221, 22 #3982.

[15] M. S. Brodskiĭ, a) *Spectral analysis of bounded linear operators with completely continuous imaginary part,* Doctoral Dissertation, Odessa, 1962. (Russian)
b) *A multiplicative representation of certain analytic operator-functions,* Dokl. Akad. Nauk SSSR 138 (1961), 751–754 = Soviet Math. Dokl. 2 (1961), 695–698. MR 24 #A1031.

[16] L. A. Sahnovič, a) *A study of the "triangular form" of nonselfadjoint operators,* Izv. Vysš. Učebn. Zaved. Mathematika 1959, no. 4 (11), 141–149; English transl., Amer. Math. Soc. Transl. (2) 54 (1966), 75–84 MR 25 #461.

b) *Dissipative operators with an absolutely continuous spectrum,* Dokl. Akad. Nauk SSSR 167 (1966), 760–763 = Soviet Math. Dokl. 7 (1966), 483–486. MR 34 #3334.

[17] N. Aronszajn and K. T. Smith, *Invariant subspaces of completely continuous operators,* Ann. of Math. (2) 60 (1954), 345–350. MR 16, 488.

[18] V. P. Potapov, *The multiplicative structure of J-contractive matrix functions,* Trudy Moskov. Mat. Obšč. 4 (1955), 125–236; English transl., Amer. Math. Soc. Transl. (2) 15 (1960), 131–243. MR 17, 958; 22 #5733.

[19] Ju. P. Ginzburg, *On J-contractive operator functions,* Dokl. Akad. Nauk SSSR 117 (1957), 171–173. (Russian) MR 20 #1203.

[20] B. Sz.-Nagy, a) *Sur les contractions de l'espace de Hilbert,* Acta Sci. Math. Szeged 15 (1953), 87–92. MR 15, 326.
b) *Sur les contractions de l'espace de Hilbert.* II, Acta Sci. Math. Szeged 18 (1957), 1–14. MR 19, 873.

[21] P. D. Lax and R. S. Phillips, *Scattering theory,* Pure and Appl. Math., vol. 26, Academic Press, New York, 1967. MR 36 #530.

[22] M. G. Kreĭn, a) *On a remarkable class of Hermitian operators,* C. R. (Dokl.) Acad. Sci. URSS 44 (1944), 175–179. MR 6, 269.
b) *The fundamental propositions of the theory of representations of Hermitian operators with deficiency index (m, m),* Ukrain. Mat. Ž. 1 (1949), no. 2, 3–66. (Russian) MR 14, 56.

[23] M. G. Kreĭn and Š. N. Saakjan, *Certain new results in the theory of resolvents of Hermitian operators,* Dokl. Akad. Nauk SSSR 169 (1966), 1269–1272 = Soviet Math. Dokl. 7 (1966), 1086–1089. MR 34 #3325.

[24] M. S. Brodskiĭ, I. C. Gohberg, M. G. Kreĭn and V. I. Macaev, *Some investigations of nonselfadjoint operators,* Proc. Fourth All-Union Math. Congress, vol. II: Sectional Lectures, "Nauka", Leningrad, 1964, pp. 261–271; English transl., Amer. Math. Soc. Transl. (2) 65 (1967), 237–251. MR 36 #3153.

[25] V. M. Adamjan and D. Z. Arov, a) *On a class of scattering operators and characteristic operator-functions of contractions,* Dokl. Akad. Nauk SSSR 160 (1965), 9–12 = Soviet Math. Dokl. 6 (1965), 1–5. MR 30 #5169.
b) *On scattering operators and contraction semigroups in Hilbert space,* Dokl. Akad. Nauk SSSR 165 (1965), 9–12 = Soviet Math. Dokl. 6 (1965), 1377–1380. MR 32 #6240.
c) *Unitary couplings of semi-unitary operators,* Akad. Nauk Armjan. SSR. Dokl. 43 (1966), no. 5, 257–263. (Russian) MR 34 #8201.

d) *Unitary couplings of semi-unitary operators,* Mat. Issled. 1 (1966), no. 2, 3–64; English transl., Amer. Math. Soc. Transl. (2) 95 (1970) (to appear). MR 34 #6528.

[26] B. Sz.-Nagy and C. Foiaş, *Analyse harmonique des opérateurs de l'espace de Hilbert,* Masson, Paris and Akad. Kiadó, Budapest, 1967. MR 37 #778.

Translated by:
A. Shuchat

Amer. Math. Soc. Transl.
(2) Vol. 90, 1970

ON MULTIPLICATIVE OPERATORS IN BANACH ALGEBRAS. I. GENERAL PROPOSITIONS*

M. S. BUDJANU AND I. C. GOHBERG

1. Let $\mathfrak{A}$ be some Banach algebra with unit element e. A closed linear operator T operating in $\mathfrak{A}$ is said to be *multiplicative* if for any pair of elements x, y of the domain ϑ_T of the operator T the following conditions are satisfied:

a) the element

$$x(Ty) + (Tx)y - \theta_T xy \in \vartheta_T, \tag{0.1}$$

where θ_T is some element of the center of the algebra $\mathfrak{A}$ not depending on x and y;

b) the relation

$$(Tx)(Ty) = T\{x(Ty) + (Tx)y - \theta_T xy\} \tag{0.2}$$

holds.

Equations with multiplicative operators of the type

$$z - T(az) = g; \quad z - T(za) = g, \tag{0.3}$$

where g and a are given elements and z is the desired element, were investigated in papers of Baxter [1] and Atkinson [2].

The history of the question and references to the literature are given in the paper [3].

2. Many problems of analysis reduce to equations of the type (0.3) with various multiplicative operators T. We present three examples.

The problem of similarity of operators. For simplicity we restrict ourselves to operators operating in finite-dimensional space. Suppose that $V = \|v_{jk}\|_1^n$ is a triangular matrix ($v_{jk} = 0$ if $j < k$). Suppose that $\Lambda = \|\lambda_j \delta_{jk}\|_1^n$ is a diagonal matrix with distinct elements ($\lambda_j \neq \lambda_k$, $j \neq k$). As is known, the matrix $A = \Lambda + V$ is similar to the matrix Λ, i.e. there exists a nonsingular matrix $X = \|x_{jk}\|_1^n$ such that

---

* Translation of Mat. Issled. 2 (1967), no. 4, 14–30.

$$X^{-1}AX = \Lambda.$$

This last equation may be rewritten in the form

$$X\Lambda - \Lambda X = VX.$$

Thus the matrix X is the solution of the matrix equation

$$\Delta X = VX, \tag{0.4}$$

where the operator Δ is defined on the set of all matrices by the equation

$$\Delta X = X\Lambda - \Lambda X.$$

Similarly to the operation of differentiation, for the operator Δ the following relation is satisfied:

$$\Delta(XY) = (\Delta X)Y + X(\Delta Y). \tag{0.5}$$

Denote by $\mathfrak{A}_n^+$ the set of all triangular matrices of the form

$$z = \|z_{jk}\|_1^n \quad (z_{jk} = 0,\ j < k).$$

Obviously the algebra $\mathfrak{A}_n^+$ is invariant relative to the operator Δ, and that operator is invertible in it. Denote the operator inverse to the restriction of Δ to $\mathfrak{A}_n^+$ by Γ. Then it follows from (0.5) that for any pair of matrices $Z_1, Z_2 \in \mathfrak{A}_n^+$

$$(\Gamma Z_1)(\Gamma Z_2) = \Gamma\{(\Gamma Z_1)Z_2 + Z_1(\Gamma Z_2)\}.$$

Thus Γ is a multiplicative operator satisfying the identity (0.2) with $\theta_\Gamma = 0$.

We shall seek the solution X of equation (0.4) in the form $X - I \in \mathfrak{A}_n^+$. Then equation (0.4) is equivalent to the following equation with multiplicative operator:

$$X - \Gamma(VX) = I. \tag{0.6}$$

The operator X^{-1} turns out to be a solution of the equation

$$Y + \Gamma(YV) = I. \tag{0.7}$$

It is easy to show that under the indicated restrictions equations (0.6) and (0.7) always have unique solutions with $Y = X^{-1}$. It is also easy to deduce a formula for these equations.

The role of equations (0.6) and (0.7) in establishing the similarity of the perturbed and unperturbed operators was apparently first discovered by Friedrichs [4, 5] (see also [6, 7]),

Problem of canonical factorization. Suppose that $\mathfrak{A}^{\pm}$ are two subalgebras of $\mathfrak{A}$ for which $\mathfrak{A}^{+} \cap \mathfrak{A}^{-} = \{0\}$.

On the linear manifold $\vartheta_P = \mathfrak{A}^{+} \cup \mathfrak{A}^{-}$ we define a projector P projecting ϑ_P onto the subalgebra $\mathfrak{A}^{+}$ parallel to $\mathfrak{A}^{-}$. It is easy to verify (see [2, 8, 9]) that P is a multiplicative operator satisfying the relation (0.2) with $\theta_P = 1$. Such projectors were called in [8] *multiplicative*.

The closed projector P in $\mathfrak{A}$ is multiplicative if and only if the subspace of values of P and the subspace of zeros of P are subalgebras of $\mathfrak{A}$. We note further that if P is a bounded multiplicative operator and $Pe = e$, then P is a multiplicative projector.

The *canonical factorization* of the element $e - a$ $(a \in \mathfrak{A})$ is (see [3]) a representation of that element in the form

$$e - a = (e + a_{-})(e + a_{+}),$$

where the elements $e + a_{\pm}$ are invertible, $a_{\pm} \in \mathfrak{A}^{\pm}$ and $(e + a_{\pm})^{-1} - e \in \mathfrak{A}^{\pm}$.

The problem of factorization of the elements $e - a$ $(a \in \mathfrak{A})$ of a Banach algebra $\mathfrak{A}$ reduces (see [3, 10]) to the solution of the following equations:

$$x - P(ax) = e \quad (a,\ x \in \mathfrak{A}),$$

$$y - (I - P)(ya) = e \quad (a,\ y \in \mathfrak{A}).$$

The factors of the factorization are completely determined by the solutions of these equations.

The Wiener-Hopf equations. Without going into details we note that the integral equations of Wiener-Hopf, systems of such equations, and their discrete analogs may be written in the form

$$x - P(ax) = y,$$

where P is a multiplicative projector.

3. In the present paper we investigate the general properties of equations of the type (0.3). Our attention will basically be devoted to the clarification of the connection between the existence and the invertibility of the solutions of the equations

$$x - T(ax) - e; \quad y - U(ya) = e,$$

where $U = \theta_T I - T$, and the invertibility of the operators $I - T_a$ and $I - U_a$, where the operators T_a and U_a are defined by the equations

$$T_a x = T(ax), \quad U_a x = U(xa) \quad (x \in \vartheta_T).$$

Our results complement some results of the papers [1, 2] and generalize some propositions of [3, 10].

§1.

1. Suppose that T is a multiplicative operator. Without difficulty one verifies that the operator $S = \alpha T$, where α is any element of the center of the algebra $\mathfrak{A}$, is also a multiplicative operator, with $\theta_S = \alpha\theta_T$.

The operator $U = \theta_T I - T$ is said to be the operator *associated with* T. It follows from (0.1) that

$$x(Uy) + (Ux)y - \theta_T xy \in \vartheta_T, \tag{1.1}$$

where $x, y \in \vartheta_T$. Moreover, one verifies directly that the operator U satisfies the relation

$$(Ux)(Uy) = U\{x(Uy) + (Ux)y - \theta_T xy\}.$$

Thus U also is a multiplicative operator, with $\theta_U = \theta_T$.

It follows from (0.1) and (1.1) that

$$x(Ty) - (Ux)y \in \vartheta_T; \quad x(Uy) - (Tx)y \in \vartheta_T.$$

If T and U are associated multiplicative operators, then for any pair of elements $x, y \in \vartheta_T$ we have the relations

$$(Tx)(Uy) = \theta_T(Tx)y + T\{x(Uy) - (Tx)y\}, \tag{1.2}$$

$$(Ux)(Ty) = \theta_T(Ux)y + U\{x(Ty) - (Ux)y\}. \tag{1.3}$$

We note also the following identities:

$$(Tx)(Uy) = \theta_T x(Uy) + U\{(Tx)y - x(Uy)\}, \tag{1.4}$$

$$(Ux)(Ty) = \theta_T x(Ty) + T\{(Ux)y - x(Ty)\}. \tag{1.5}$$

2. **Lemma 1.1** [2]. *Suppose that T is a multiplicative operator, U the operator associated to it, and a some element of $\mathfrak{A}$. If the elements u and v satisfy the equations*

$$u - T(au) = e, \tag{1.6}$$

$$v - U(va) = e, \tag{1.7}$$

then

$$v(e - \theta_T a)u = e. \tag{1.8}$$

Proof. (1.6) and (1.7) imply the relation

$$vu = e + T(au) + U(va) + \{U(va)\}\{T(au)\}.$$

Using the identity (1.3), we get

$$\begin{aligned} vu &= e + T(au) + U(va) + \theta_T \{U(va)\} au \\ &\quad + U\{va[T(au)] - [U(va)]au\} \\ &= e + T(au) + \theta_T\{U(va)\}au + U\{va(u-e) - (v-e)au + va\} \\ &= e + T(au) + U(au) + \theta_T\{U(va)\}au. \end{aligned}$$

Since $T(au) + U(au) = \theta_T au$, it follows from the last equation that

$$vu = e + \theta_T au + \theta_T (v - e) au = e + \theta_T vau.$$

The resulting equality is equivalent to (1.8).

We note that if $\mathfrak{A}$ is an algebra with invertible divisors of unity and if u and v are solutions of equations (1.6) and (1.7) with associated multiplicative operators, then it follows from (1.8) that the solutions u and v and the element $e - \theta_T a$ are invertible in $\mathfrak{A}$.

In generalizing the definition of a canonical factorization we may introduce a canonical factorization relative to any multiplicative operator T. In fact it is natural to call *canonical factorization of the element* $e - \theta_T a$ any representation of it in the form

$$e - \theta_T a = v^{-1} u^{-1} \quad (a \in \mathfrak{A}),$$

where u and v are solutions of equations (1.6), (1.7). We recall that here we are assuming that equations (1.6) and (1.7) are solvable and that their solutions are invertible.

In the case when the operator T is a multiplicative projector, i.e. $\mathfrak{A}^+ = T\mathfrak{A}$, $\mathfrak{A}^- = (I - T)\mathfrak{A}$, this definition is equivalent to the one given above.

3. To each multiplicative operator T we may associate the set $\mathfrak{M}_T$ of all elements $a \in \vartheta_T$ for which $xay \in \vartheta_T$ for any $x, y \in \mathfrak{A}$.

It is easy to see that $\mathfrak{M}_T$ *is a maximal two-sided ideal of the algebra* $\mathfrak{A}$, *lying in* ϑ_T. *If* T *and* U *are associated multiplicative operators, then obviously* $\mathfrak{M}_T = \mathfrak{M}_U$.

Theorem 1.1. *Suppose that* T *is a multiplicative operator,* U *the operator associated to it, and* a *some element of* $\mathfrak{M}_T$.

a) *If equation* (1.6) *is solvable and if for its solution* u *the element* $(e - \theta_T a)u$ *has a right inverse* $\{(e - \theta_T a)u\}_r^{-1}$, *then the operator* $I - T_a$ $(T_a x = T(ax);\ x \in \mathfrak{A})$ *is invertible on the right, its right inverse being defined by the equation*

$$(I - T_a)_r^{-1} f = f + uT\{[(e - \theta_T a) u]_r^{-1} af\}. \tag{1.9}$$

b) *If equation* (1.7) *is solvable and if for its solution* v *the element* $v(e - \theta_T a)$ *has a left inverse, then the operator* $I - T_a$ *is left invertible, the left inverse operator being given by*

$$(I - T_a)_l^{-1} f = f + [v(e - \theta_T a)]_l^{-1} T(vaf). \tag{1.10}$$

c) *If both equations* (1.6) *and* (1.7) *are solvable and if their solutions* u *and* v *are invertible, then the operator* $I - T_a$ *is invertible, with*

$$(I - T_a)^{-1} f = f + uT(vaf) \quad (f \in \mathfrak{A}). \tag{1.11}$$

Proof. Since the equation

$$x - T(ax) = f \quad (a \in \mathfrak{M}_T)$$

may be written in the form

$$x - f - T\{a(x - f)\} = T(af),$$

assertion a) will be proved as soon as one shows that the element

$$z = uTw, \text{ where } w = \{(e - \theta_T a) u\}_r^{-1} af,$$

is a solution of the equation

$$z - T(az) = T(af) \tag{1.12}$$

for any $f \in \mathfrak{A}$. Denote by J the expression

$$J = uTw - T(auTw).$$

In view of identity (0.2), for any elements au, w $(\in \vartheta_T)$ we will have

$$[T(au)][Tw] = T\{auTw + [T(au)]w - \theta_T auw\}$$

and therefore

$$J = uTw - [T(au)][Tw] + T\{[T(au)]w - \theta_T auw\}.$$

Taking account of the fact that u is a solution of (1.6), we get

$$J = uTw - (u - e)Tw + T\{(u - e)w - \theta_T auw\}$$
$$= T\{(e - \theta_T a) uw\}.$$

Replacing w here by its value $[(e - \theta_T a) u]_r^{-1} af$, we arrive at the equation $J = T(af)$.

In order to prove assertion b) it is sufficient to prove that if equation (1.12) is solvable, then its solution z has the form

$$z = [v(e - \theta_T a)]_l^{-1} T(vaf). \tag{1.13}$$

(1.7) and (1.2) imply the relation

$$vz = T\{a(f+z)\} + \{U(va)\} \cdot \{T[a(f+z)]\}.$$

Hence, using identity (1.5), we get

$$\begin{aligned} vz &= T[a(f+z)] + \theta_T vaT[a(f+z)] \\ &\quad + T\{[U(va)]a(f+z) - va[T(a(f+z))]\} \\ &= \theta_T vaz + T\{[e + U(va)]a(f+z) - va[T(a(f+z))]\} \\ &= \theta_T vaz + T[va(f+z) - vaz] = \theta_T vaz + T(vaf), \end{aligned}$$

from which (1.13) follows.

Finally, assertion c) follows directly from a) and b) on taking account of Lemma 1.1.

This theorem directly implies the following.

Corollary 1.1. *If the hypotheses of assertion* a) *are satisfied, then the operator* $I - U_a$ $(U_a x = U(xa))$ *is left invertible, with*

$$(I - U_a)_l^{-1}\phi = \phi + U(\phi au)[(e - \theta_T a)u]_r^{-1}.$$

If the hypotheses of assertion b) *are satisfied, then the operator* $I - U_a$ *is right invertible, with*

$$(I - U_a)_r^{-1}\phi = \phi + U\{\phi a\{v(e - \theta_r a)_l^{-1}\}v.$$

If now the hypotheses of assertion c) *are satisfied, then the operator* $I - U_a$ *is invertible, with*

$$(I - U_a)^{-1}\phi = \phi + U(\phi au)v.$$

This corollary follows from Theorem 1.1 by introducing the new multiplication $x \circ y = yx$ into the algebra $\mathfrak{A}$.

Corollary 1.2. *Suppose that* $\mathfrak{A}$ *is an algebra in which all the divisors of unity are invertible. For the operators* $I - T_a$, $I - U_a$ *to be invertible, it is necessary and sufficient that equations* (1.6) *and* (1.7) *be solvable.*

Assertions close to Theorem 1.1 were proved in [2] under additional restrictions. For the case when $a = e$, formula (1.11) is contained in [11]. Theorem 1.1 also generalizes some propositions in the papers [3, 10].

4. **Theorem 1.2.** *Suppose that* T *is a multiplicative operator and that* $a \in \mathfrak{M}_T$. *Suppose the following conditions are satisfied:*

a) *the equation*

$$u - T(au) = e \tag{1.14}$$

has a solution u *for which the element* $(e - \theta_T a)u$ *has a right inverse* $w = [(e - \theta_T a)u]_r^{-1}$;

b) *for some element* $b \in \mathfrak{M}_T$ *the equation*

$$v - T(wbuv) = e \tag{1.15}$$

has a solution v.

Then the element $z = uv$ *is a solution of the equation*

$$z - T\{(a+b)z\} = e.$$

Proof. Multiplying equations (1.14) and (1.15) and using the identity (0.2), we get

$$uv = e + T(au) + T(wbuv)$$
$$+ T\{auT(wbuv) + [T(au)]wbuv - \theta_T auwbuv\}$$
$$= e + T\{au[e + T(wbuv)] + [e + T(au)]wbuv - \theta_T auwbuv\}.$$

In view of (1.14) and (1.15) it therefore follows that

$$uv = e + T\{auv + (e - \theta_T a)uwbuv\}.$$

Accordingly,

$$uv = e + T\{(a+b)uv\}.$$

Corollary 1.3. *Let conditions* a) *and* b) *of Theorem* 1.2 *be replaced by the conditions*:

a°) *the equation*

$$u - T(ua) = e$$

has a solution u *for which the element* $u(e - \theta_T a)$ *has a left inverse* $w = [u(e - \theta_T a)]_l^{-1}$;

b°) *for some element* $b \in \mathfrak{M}_T$ *the equation*

$$v - T(vubw) = e$$

has a solution v.

Then the conclusion of Theorem 1.1 *may be replaced by the assertion that the element* $z = vu$ *is a solution of the equation*

$$z - T\{z(a+b)\} = e.$$

Theorem 1.3 (Multiplication theorem). *Suppose that* $\mathfrak{A}$ *is a commutative Banach algebra,* T *a multiplicative operator,* a *and* b *elements of* $\mathfrak{M}_T$, *and the element* $c \in \mathfrak{M}_T$ *is such that*

$$c = a + b - \theta_T ab. \tag{1.16}$$

If the elements u, z *of* $\mathfrak{A}$ *are solutions of the equations*

$$u - T(au) = e, \tag{1.17}$$

$$z - T(bz) = e, \tag{1.18}$$

then the element $w = uz$ *is a solution of the equation*

$$w - T(cw) = e.$$

Proof. It follows from (1.17) and (1.18) that

$$uz = e + T(au) + T(bz) + [T(au)][T(bz)].$$

From this last equation, using (0.2), we get

$$\begin{aligned} uz &= e + T\{au + bz + auT(bz) + [T(au)]bz - \theta_T aubz\} \\ &= e + T(auz + ubz - \theta_T aubz). \end{aligned}$$

Since the algebra $\mathfrak{A}$ is commutative,

$$uz = e + T\{(a + b - \theta_T ab)uz\},$$

which proves the theorem.

We note that if θ_T is an invertible element in $\mathfrak{A}$, then condition (1.6) is equivalent to the condition

$$(e - \theta_T a)(e - \theta_T b) = e - \theta_T c.$$

5. In some cases (see for example [4]) one considers equations of the form

$$y - aTy = f \tag{1.19}$$

with a multiplicative operator T, with given elements a and f, and with y the unknown.

Equation (1.19) is closely connected with the equations considered above. One verifies easily that the general solution of equation (1.19) has the form

$$y = f + ax,$$

where x is the general solution of the equation

$$x - T(ax) = Tf.$$

On the other hand, one also easily verifies that the general solution of the equation

$$x - T(ax) = f$$

has the form

$$x = f + Ty,$$

where y is the general solution of the equation

$$y - aTy = af.$$

§2.

1. Suppose first that T is a multiplicative operator and that $\mathfrak{M}_T$ is the set of all elements $a \in \vartheta_T$ for which $xay \in \vartheta_T$ for any $x, y \in \mathfrak{A}$. As we already noted, $\mathfrak{M}_T$ is a maximal two-sided ideal of the algebra $\mathfrak{A}$, lying in ϑ_T.

Lemma 2.1. *The ideal $\mathfrak{M}_T$ is a symmetrically-normed ideal of the algebra $\mathfrak{A}$ with the norm*

$$\|a\| = \max\{|a|, \sup|T(xay)|, \sup|U(xay)|\}.$$

This lemma is proved in the same way as Theorem 1.1 of [10], where the definition of symmetrically-normed ideal will be found.

For elements $a \in \mathfrak{M}_T$ the operators T_a and U_a are bounded, with

$$|T_a| \leq \|a\|, \; |U_a| \leq \|a\|.$$

Denote by Λ_T the set of all elements $a \in \mathfrak{M}_T$ for which $I - T_a$ and $I - U_a$ are invertible. Obviously the set Λ_T is an open set in $\mathfrak{M}_T$, in the sense of the metric of $\mathfrak{M}_T$.

Denote by Φ_T the set of elements $a \in \mathfrak{M}_T$ for which the equations

$$x - T(ax) = e, \; y - U(ya) = e \tag{2.1}$$

have invertible solutions. It follows from Theorem 1.1 that $\Phi_T \subset \Lambda_T$.

Theorem 2.1. *The set Φ_T is an open set in $\mathfrak{M}_T$. If Λ_a is a connected component of Λ_T, then either $\Lambda_a \subset \Phi_T$, or else the intersection of Λ_a and Φ_T is empty. The connected components Λ_0 and Φ_0 of the sets Λ_T and Φ_T containing the element $a = 0$ coincide.*

Proof. Suppose that $a \in \Phi_T$. Then from Lemma 1.1 the element $e - \theta_T a$ is invertible. For any element $\Delta a \in \mathfrak{M}_T$ we consider the equations

$$x_1 - T(w_1 \Delta a x_1) = e, \tag{2.2}$$

$$y_1 - U(y_1 y \Delta a w_2) = e, \tag{2.3}$$

where

$$w_1 = \{(e - \theta_T a)x\}^{-1}, \quad w_2 = \{y(e - \theta_T a)\}^{-1},$$

and x and y are invertible solutions of equations (2.1).

It is easy to see that if Δa has a sufficiently small norm $\|\Delta a\|$, then equations (2.2), (2.3) have solutions x_1, y_1 which are inverse elements in the algebra $\mathfrak{A}$.

According to Theorem 1.2 the elements $\tilde{x} = xx_1$, $\tilde{y} = y_1 y$ are solutions of the equations

$$\tilde{x} - T\{(a + \Delta a)\tilde{x}\} = e, \tag{2.4}$$

$$\tilde{y} - U\{\tilde{y}(a + \Delta a)\} = e. \tag{2.5}$$

Obviously the elements $\tilde{x}$, $\tilde{y}$ are invertible. Accordingly $a + \Delta a \in \Phi_T$. Thus Φ_T is an open set.

Now we turn to the proof of the second part of the theorem. Suppose that equations (2.1) are solvable for each $a \in \Lambda_\alpha$, that for $a_0 \in \Lambda_\alpha$ their solutions x_0 and y_0 are invertible, and suppose that $a(t)$ $(0 \leq t \leq 1)$ is a continuous vector-function on $\mathfrak{M}_T$ with values from Λ_α, joining the element a_0 with the element $a \in \Lambda_\alpha$: $a(0) = a_0$, $a(1) = a$.

Consider the vector functions

$$x(t) = (I - T_{a(t)})^{-1} e \quad (0 \leq t \leq 1), \tag{2.6}$$

$$y(t) = (I - U_{a(t)})^{-1} e \quad (0 \leq t \leq 1). \tag{2.7}$$

They are obviously continuous, and their values at the point $t = 0$ are inverse elements in $\mathfrak{A}$. From Lemma 1.1 it follows from (2.6) and (2.7) that

$$y(t)(e - \theta_T a(t))x(t) = e \quad (0 \leq t \leq 1). \tag{2.8}$$

It therefore follows that at each point t $(0 \leq t \leq 1)$ the elements $x(t)$ and $y(t)$ are invertible on the left and right respectively. Since for $t = 0$ these elements are invertible on both sides, in view of the auxiliary proposition on p. 202 of [12] (transl. p. 167), $x(t)$ and $y(t)$ are also invertible on both sides for all t $(0 \leq t \leq 1)$, from which it follows that $a \in \Phi_T$.

The theorem is proved.

2. Denote by $\Lambda_T(a)$ $(a \in \mathfrak{M}_T)$ the set of complex numbers λ for which the operators $I - \lambda T_a$ and $I - \lambda U_a$ are invertible, and by $\Phi_T(a)$ the set of

numbers λ for which $\lambda a \in \Phi_T$. It is easily deduced from Theorem 2.1 that the set $\Phi_T(a)$ either coincides with $\Lambda_T(a)$ or consists of the union of connected components of $\Lambda_T(a)$. Since the point $\lambda = 0 \in \Lambda_T(a) \cap \Phi_T(a)$, the connected components of $\Lambda_T(a)$ and $\Phi_T(a)$ containing the point $\lambda = 0$ coincide. In particular, for the radius $r(a)$ of the largest circle with center at the point $\lambda = 0$ and contained in $\Phi_T(a)$, the following equation holds:

$$r(a) = \min(\rho(T_a), \rho(U_a)),$$

where

$$\rho(T_a) = \left[\lim_{n\to\infty} \sqrt[n]{|T_a^n|}\right]^{-1},$$

$$\rho(U_a) = \left[\lim_{n\to\infty} \sqrt[n]{|U_a^n|}\right]^{-1}.$$

BIBLIOGRAPHY

[1] G. Baxter, *An analytic problem whose solution follows from a single algebraic identity*, Pacific J. Math. **10** (1960), 731–742. MR 22 #9990.

[2] F. V. Atkinson, *Some aspects of Baxter's functional equation*, J. Math. Anal. Appl. 7 (1963), 1–30. MR 27 #5135.

[3] M. S. Budjanu and I. C. Gohberg, *The factorization problem in abstract Banach algebras*. I: *Splitting algebras*, Mat. Issled. 2 (1967), no. 2, 25–51. (Russian) MR **37** #5697.

[4] K. O. Friedrichs, *On the perturbation of continuous spectra*, Comm. Pure Appl. Math. **1** (1948), 361–406. MR **10**, 547.

[5] ———, *Spectral theory of operators in Hilbert space*, Courant Institute of Math. Sci., New York, 1960.

[6] J. M. Freeman, *Perturbations of the shift operator*, Trans. Amer. Math. Soc. **114** (1965), 251–260. MR **30** #2342.

[7] T. Kato, *Perturbation theory for linear operators*, Die Grundlehren der math. Wissenschaften, Band 132, Springer-Verlag, New York, 1966. MR **34** #3324.

[8] I. C. Gohberg, *A generalization of theorems of M. G. Kreĭn of the type of the Wiener-Levi theorems*, Mat. Issled. **1** (1966), no. 1, 110–130. (Russian) MR **34** #3366.

[9] M. A. Barkar′, *On functions of operators*, Mat. Issled. 2 (1967), no. 1, 110–116. (Russian) MR 35 #2160.

[10] M. S. Budjanu and I. C. Gohberg, *The factorization problem in abstract Banach algebras*. II: *Irreducible algebras*, Mat. Issled. 2 (1967), no. 3, 3–19. (Russian) MR **37** #5698.

[11] J. B. Miller, *Some properties of Baxter operators*, Acta Math. Acad. Sci. Hungar. **17** (1966), 387–400. MR **34** #4909.

[12] I. C. Gohberg and M. G. Kreĭn, *Theory and applications of Volterra operators in Hilbert space*, "Nauka", Moscow, 1967; English transl., Transl. Math. Monographs, vol. 24, Amer. Math. Soc., Providence, R. I., 1970. MR **36** #2007.

Translated by:
J. M. Danskin

Amer. Math. Soc. Transl.
(2) Vol 90, 1970

THE CHARACTERISTIC OPERATOR-FUNCTION OF AN ARBITRARY BOUNDED OPERATOR*

UDC 517.432

O. V. KUŽEL'

1. In the investigation of operators differing "a little" from unitary operators, several authors [1-6] have introduced the concept of characteristic matrix-functions, the investigation of which makes it possible to obtain certain properties of the operators under consideration (in particular: to examine the spectrum and invariant subspaces, to build the "triangular model", which is the infinite-dimensional analog of the famous theorem on bringing a finite matrix to triangular form, etc.). An operator T is said to differ "a little" from unitary if the operator $I-T^*T$ is completely continuous, i.e. compact (in particular, if it is finite dimensional).

In [7] (and also [8]) the last condition (the complete continuity of the operator $I-T^*T$) was rejected for the contraction operators ($\|T\| < 1$) and the characteristic operator-function was constructed and investigated for an arbitrary contraction operator. In connection with this there arises the question of whether it is possible to reject the constraint $\|T\| < 1$. It turns out that the answer is in the affirmative.

In the present paper there will be given a universal definition of the characteristic operator-function of an arbitrary bounded operator, and we shall examine the properties of this operator-function. As a special case we shall obtain an expression for the characteristic operator-function of a contraction operator and we shall establish the connection between the characteristic operator-function and characteristic matrix-function of K_r-operators [6].

2. Let T be a bounded linear operator, acting in a Hilbert space H and with $D_T = H$ (D_T is the domain of the operator T). Put $Q_T = I-T^*T$ and denote by G_T the kernel of the operator Q_T (G_T is the set of those vectors f from H which are mapped into zero by Q_T). It is easy to show that G_T is a subspace of H, and also $TG_T \subset G_{T^*}$, $T^*G_{T^*} \subset G_T$ and $\mathfrak{N}_T = H \ominus G_T$ is the closure of the range (set of values) of the operator Q_T.

Moreover, let

* Translation of Dopovīdī Akad. Nauk Ukraīn. RSR Ser. A 1968, 233–235.

$$Q_T = J|Q_T| \quad (|Q_T| = \sqrt{Q_T^2}) \tag{1}$$

be the polar representation of the operator Q_T. In expression (1) J is a partially isometric [9] Hermitian operator, connected with the projection operator into the subspace $\mathfrak{N}_T$ by the relation $J^2 = P$.

Definition. The operator-function $\Theta_T(z)$ which is defined for any z ($\bar{z}^{-1} \in \rho(T)$, where $\rho(T)$ is the resolvent set of the operator T) by the relation

$$\Theta_T(z) = TJ - z|Q_{T^*}|^{\frac{1}{2}}(I - zT^*)^{-1}|Q_T|^{\frac{1}{2}}, \tag{2}$$

is called the *characteristic operator-function* of the operator T.

In particular, if $||T|| < 1$ then $|Q_T| = Q_T$ and the operator J coincides with the projection operator P. Therefore in this case relation (2) can be rewritten in the form

$$\Theta_T(z) = [T - zQ_{T^*}^{\frac{1}{2}}(I - zT^*)^{-1}Q_T^{\frac{1}{2}}]|_{\mathfrak{N}_T}.$$

Consequently we have obtained, as a special case, the well-known expression for the characteristic operator-function of a contraction operator

By relation (2) the characteristic operator-function maps the space H into the subspace $\mathfrak{N}_{T^*}$ and is connected with the characteristic operator-function of the operator T^* by the relation

$$\Theta_T^*(z) = \Theta_{T^*}(\bar{z}). \tag{3}$$

Moreover, if z and λ are such that $\bar{z}^{-1} \in \rho(T)$ and $\bar{\lambda}^{-1} \in \rho(T)$, then

$$\tilde{J} - \Theta_T(z) J\Theta_T(\lambda) = (1 - z\bar{\lambda})\, T_z T_\lambda^*, \tag{4}$$

where

$$T_z = |Q_{T^*}|^{\frac{1}{2}}(I - zT^*)^{-1}, \tag{5}$$

and the operator $\tilde{J}$ is taken from the polar representation of the operator Q_{T^*} ($Q_{T^*} = \tilde{J}|Q_{T^*}|$). For $z = \bar{\lambda}$ and $\lambda = z^{-1}$ relation (4) implies the equality

$$\Theta_T(\bar{\lambda})\, J\Theta_T^*\left(\frac{1}{\lambda}\right) = \tilde{J}. \tag{6}$$

If we put $z = \lambda$ in (4), we arrive at the following important inequalities:

$$\Theta_T(\lambda)\, J\Theta_T^*(\lambda) \leqslant \tilde{J} \quad (|\lambda| \leqslant 1),$$

$$\Theta_T(\lambda)\, J\Theta_T^*(\lambda) \geqslant \tilde{J} \quad (|\lambda| \geqslant 1).$$

Analogously it can be established that

$$J - \Theta_T^*(\lambda)\, \tilde{J}\Theta_T(z) = (1 - \bar{\lambda}z)|Q_T|^{\frac{1}{2}}(I - \bar{\lambda}T)^{-1}(I - zT^*)^{-1}|Q_T|^{\frac{1}{2}},$$

whence follow these relations:

$$\Theta_T^*(\lambda)\tilde{J}\Theta_T\left(\frac{1}{\bar{\lambda}}\right) = J,$$

$$\Theta_T^*(\lambda)\tilde{J}\Theta_T(\lambda) \leqslant J \quad (|\lambda| < 1),$$

$$\Theta_T^*(\lambda)\tilde{J}\Theta_T(\lambda) \geqslant J \quad (|\lambda| \geqslant 1).$$

Note also that the characteristic operator-function satisfies these relations:

$$\Theta_T(z)\, J\Theta_T^*(0) = \tilde{J} - |Q_{T^*}|^{\frac{1}{2}}(I - zT^*)^{-1}|Q_{T^*}|^{\frac{1}{2}}, \tag{7}$$

$$\Theta_T^*(0)\tilde{J}\Theta_T(z) = J - |Q_T|^{\frac{1}{2}}(I - zT^*)^{-1}|Q_T|^{\frac{1}{2}}. \tag{8}$$

3. Now let $\dim \mathfrak{N}_T = r < \infty$ and suppose the operator T has a continuous inverse. Consider an orthonormal basis $\{e_k\}_{k=1}^r$ of the subspace $\mathfrak{N}_T$, consisting of the eigenvectors of the operator Q_T. Let

$$\tilde{e}_k = \frac{1}{\sqrt{1-\lambda_k}}\, Te_k \quad (k = 1, 2, \ldots, r),$$

where λ_k is the eigenvalue of the operator Q_T belonging to e_k. The vectors $\{\tilde{e}_k\}_{k=1}^r$ form an orthonormal basis of the subspace $\mathfrak{N}_{T^*}$ and coincide with the eigenvectors of the operator Q_{T^*}. Consider the matrix function

$$\chi_T(z) = \|(\Theta_T(z)\, e_k, \tilde{e}_i)\|.$$

Making use of the properties of the operators J, $\tilde{J}$ and $|Q_T|^{1/2}$, and relation (8), we obtain that $\chi_T(z)$ satisfies the equation

$$\chi_T(z)\, J\chi_T^*(0) = J - \|(I - zT^*)^{-1}h_k, h_i)\|,$$

where $h_k = |\lambda_k|^{1/2}e_k$ $(k = 1, 2, \cdots, r)$ is the α-basis of the operator T, and $J = \|J_k\delta_{ki}\|$ $(J_k = \operatorname{sign}\lambda_k)$ is the matrix of coefficients. Thus, by what was said above, $\chi_T(z)$ is the characteristic matrix-function [6] of the operator T.

BIBLIOGRAPHY

[1] M. S. Livšic, *On a certain class of operators in a Hilbert space*, Mat. Sb. 19 (61) (1946), 239–262; English transl., Amer. Math. Soc. Transl. (2) 13 (1960), 61–83. MR 8, 588; 22 #3981a.

[2] ———, *Isometric operators with equal deficiency indices, quasi-unitary operators*, Mat. Sb. 26 (68) (1950), 247–264; English transl., Amer. Math. Soc. Transl. (2) 13 (1960), 85–103. MR 11, 669; 22 #3981b.

[3] M. S. Livšic and V. P. Potapov, *A theorem on the multiplication of characteristic matrix functions,* Dokl. Akad. Nauk SSSR 72 (1950), 625–628. (Russian) MR 11, 669.

[4] V. T. Poljackiĭ, *On the reduction of quasi-unitary operators to a triangular form,* Dokl. Akad. Nauk SSSR 113 (1957), 756–759. (Russian) MR 19, 873.

[5] A. V. Kužel', *Multiplication theorem for characteristic matrix functions of non-unitary operators,* Naučn. Dokl. Vysš. Školy. Fiz.-Mat. Nauk 1959, no. 3, 33–41. (Russian)

[6] ———, *Characteristic matrix functions of quasi-unitary operators of arbitrary rank in a space with indefinite metric,* Dopovīdī Akad. Nauk Ukraïn. RSR 1962, 1135–1138. (Ukrainian) MR 26 #1761.

[7] B. Sz.-Nagy and C. Foiaş, *Sur les contractions de l'espace de Hilbert.* VIII: *Fonctions caractéristiques. Modèles fonctionnels,* Acta Sci. Math. (Szeged) 25 (1964), 38–71. MR 30 #2348.

[8] I. C. Gohberg and M. G. Kreĭn, *On a description of contraction operators similar to unitary ones,* Funkcional. Anal. i Priložen. 1 (1967), 38–60 = Functional Anal. Appl. 1 (1967), 33–52. MR 35 #4761.

[9] ———, *Introduction to the theory of linear non-selfadjoint operators in Hilbert space,* "Nauka", Moscow, 1965; English transl., Transl. Math. Monographs, vol. 18, Amer. Math. Soc., Providence, R. I., 1969. MR 36 #3137.

Translated by:
B. Bollabas

Amer. Math. Soc. Transl.
(2) Vol. 90, 1970

ON ESTIMATION OF THE SPECTRUM OF A CLASS OF POSITIVE LINEAR OPERATORS*

UDC 517.43+513.881

I. A. BAHTIN

In [1] M. G. Kreĭn showed the existence of a common eigenvector for the class $\{A^*\}$ of conjugate operators under the assumption that the commutative class $\{A\}$ of positive linear operators A maps the interior of a solid semigroup K into itself.

In this connection there arises the question of extending the results of Kreĭn to the case when the semigroup K is not solid.

Here we give a positive answer to this question. It turns out that the solution is closely related to conditions of continuity of an additive functional. This, in turn, gives ground for extending the work of M. A. Krasnosel'skiĭ, V. Ja. Stecenko and I. A. Bahtin [2] on the continuity of additive positive operators to the case of a generating semigroup K.

We use the terminology of the theory of cones in a Banach space (see [1]).

§1. Continuity of additive positive operators

1. We first introduce some definitions and notation. Let E be a real Banach space. A set $K \subset E$ is called a *linear semigroup* [1] if it is such that:

if $x \in K$ then $\lambda x \in K$ for all $\lambda \geq 0$,

if $x, y \in K$ then $x + y \in K$.

Any semigroup K allows the introduction of a partial order on E: we write $x \leqslant y$ or $y \geqslant x$ if $y - x \in K$. The relation $x \leqslant y$ has many properties in common with scalar inequality:

if $x \leqslant y$, $y \leqslant z$, then $x \leqslant z$;

if $x \leqslant y$, then $y \leqslant -x$;

if $x \leqslant y$, $\lambda > 0$, then $\lambda x \leqslant \lambda y$;

if $x_1 \leqslant y_1$, $x_2 \leqslant y_2$ then $x_1 + x_2 \leqslant y_1 + y_2$.

A closed semigroup K is called a *cone* [1] if $x \in K (x \neq \theta,\ \theta =$ the zero of $E)$ implies $-x \notin K$.

K is called *solid* if it contains an interior point. It is called *generating*

---

* Translation of Mat. Sb. 67 (109) (1965), 267–278

if each $x \in E$ can be written in the form $x = u - v$ with $u, v \in K$.

An operator A on E is called *positive* if $AK \subset K$. A is called *monotone* if $x, y \in E$ and $x \leqslant y$ imply $Ax \leqslant Ay$.

Clearly a positive linear operator A is monotone. A is called u_0*-bounded* on E ($u_0 \in K$, $u_0 \neq \theta$) if for any $x \in E$ there exist a natural number $n = n(x)$ and a scalar $\beta = \beta(x) > 0$ such that

$$A^n x \leqslant \beta u_0. \tag{1}$$

It is easy to see that if a positive linear operator A is u_0-bounded (above) [3] on a generating semigroup K, then A is u_0-bounded also on E.

2. **Theorem 1.** *If a positive linear operator A is u_0-bounded above on a closed semigroup K, then there exist a natural number m_0 and a scalar $\beta > 0$, such that for any $x \in K$*

$$A^{n_0} x \leqslant \beta \| x \| u_0. \tag{2}$$

Proof. Let $K_{m,n}$ be the set of all $x \in K$ such that $A^n x \leqslant m \|x\| u_0$.

Clearly $K = \bigcup_{m,n=1}^{\infty} K_{m,n}$. From the fact that as a complete metric space K is of second category it follows that at least one of the sets $K_{m,n}$ is not nowhere dense in K. Hence there exists a sphere $T(x_0, \rho)$: $\|x - x_0\| \leq \rho$ with center at x_0 such that some set $K_{m_0 n_0}$ is everywhere dense in $K \cap T(x_0, \rho)$.

Consider the rings (i.e. annuli) $T(x_0, r, \rho)$ and $T(\theta, r, \rho)$: $r \leq \|x - x_0\| \leq \rho$ and $r \leq \|x\| \leq \rho$, where $0 < r < \rho$.

Clearly the set $\Pi_0 = K \cap T(\theta, r, \rho)$ is contained in the set $\Pi_1 = (K \cap T(x_0, r, \rho)) \ominus x_0$, and for any $z = x - x_0 \in \Pi_2 = (K_{m_0 n_0} \cap T(x_0, r, \rho)) \ominus x_0$ we have

$$A^{n_0} z = A^{n_0} x - A^{n_0} x_0 \leqslant A^{n_0} x \leqslant m_0 \| x \| u_0 \leqslant \frac{m_0 (\rho + \| x_0 \|)}{r} \| z \| u_0 = \beta \| z \| u_0.$$

From the density of Π_2 in Π_1, the continuity of A implies that for all $x \in \Pi_0$

$$A^{n_0} x \leqslant \beta \| x \| u_0. \tag{3}$$

Since together with any $x \in \Pi_0$ inequality (3) is also satisfied by any $y = tx$, $t \geq 0$, (3) is valid for all $x \in K$. The theorem is proved.

Note that if the positive linear operator A is u_0-bounded below [3] on the closed semigroup K, i.e. if for all $x \in K$ there exist a natural number

$n = n(x)$ and a scalar $\alpha = \alpha(x) > 0$ such that $A^n x \gg \alpha u_0$, then there still may not be a universal n_0 such that for any $x \in K (x \neq \theta)$ there exists an $\alpha = \alpha(x) > 0$ such that $A^{n_0} x \gg \alpha u_0$.

Thus, for example, in the space m of all bounded sequences $x = \{\xi_1, \xi_2, \cdots, \xi_n, \cdots\}$ with cone consisting of x's with nonnegative components then the operator $Ax = y + \xi_1 u_0$, where $y = \{\xi_2, \xi_3, \cdots, \xi_{n+1}, \cdots\}$ and $u_0 = \{1, 1, \cdots, 1, \cdots, \}$, is bounded below. However, there is no universal n_0.

3. Below we do a theorem of M. A. Krasnosel'skiĭ [3] for the case of a cone. Our proof with the assumption of the continuty of A differs from the proof of Krasnosel'skiĭ.

Definition. A positive operator A is called u_0-*positive* [3] on the semigroup K if for any nonzero $x \in K$ there exist a natural number n and scalars $\alpha > 0$, $\beta > 0$ such that $\alpha u_0 \leqslant A^n x \leqslant \beta u_0$.

Theorem 2. *If a linear operator A is positive on a closed semigroup K and if for any nonzero $x \in K$ there exist natural numbers n_1 and n_2 and scalars $\alpha, \beta > 0$ such that*

$$\alpha u_0 \leqslant A^{n_1} x, \qquad A^{n_2} x \leqslant \beta u_0,$$

then A is u_0-positive on K.

Proof. Note that for some K we have

$$\alpha_0 u_0 \leqslant A^k u_0 \leqslant \beta_0 u_0. \tag{4}$$

For k, for example, take the product $k = st$, where

$$\alpha_1 u_0 \leqslant A^s u_0, \quad A^t u_0 \leqslant \beta_1 u_0.$$

Then (4) holds for $\alpha_0 = \alpha_1^t$ and $\beta_0 = \beta_1^s$.

Since the conditions of Theorem 1 hold for A, there exists a universal n_0 such that for any $x \in K$ we have for some $\beta = \beta(x) > 0$

$$A^{n_0} x \leqslant \beta u_0.$$

For a nonzero $x \in K$ let

$$\alpha u_0 \leqslant A^{n_1} x \quad (\alpha > 0)$$

hold. Then for $n = n_0 k + n_1$

$$\alpha_2 u_0 \leqslant A^n x = A^{n_0 k} (A^{n_1} x) = A^{n_0} (A^{n_0(k-1)+n_1} x) \leqslant \beta_2 u_0,$$

where $\alpha_2 = \alpha\alpha_0^{n_0}$ and $\beta_2 > 0$.

The theorem is proved.

Note that Theorem 2 is valid even without the continuity of A.

4. Let us prove the following assertion analogous to Theorem 1.

Theorem 3. *If a positive linear operator A on a closed semigroup K is u_0-bounded on E, then there exist a natural number n_0 and a scalar $\beta_0 > 0$ such that for any $x \in E$*

$$-\beta_0 \|x\| u_0 \leqslant A^{n_0}x \leqslant \beta_0 \|x\| u_0. \tag{5}$$

Proof. First we prove that for any $x \in E$ there exist a natural number n and scalar $\beta > 0$ such that

$$-\beta u_0 \leqslant A^n x \leqslant \beta u_0.$$

In fact, the conditions of Theorem 1 hold for A. Hence there exists a universal number p_0 such that for any $y \in K$ and for some $\gamma = \gamma(y) > 0$ we have $A^{p_0} y \leqslant \gamma u_0$.

Moreover, for any $x \in E$ and for some n_1 and n_2

$$A^{n_1}x \leqslant \beta_1 u_0, \qquad -A^{n_2}x \leqslant \beta_2 u_0,$$

where $\beta_1, \beta_2 > 0$. Hence for $n = n_1 + n_2 + p_0$ and some $\beta > 0$

$$-\beta u_0 \leqslant -\beta_2 A^{p_0}(A^{n_1}u_0) = -\beta_2 A^{n_1+p_0}u_0 \leqslant A^n x \leqslant \beta_1 A^{n_2+p_0}u_0 = \beta_1 A^{p_0}(A^{n_2}u_0) \leqslant \beta u_0.$$

Let $E_{m,n}$ be the set of all $x \in E$ such that

$$-m\|x\| u_0 \leqslant A^n x \leqslant m \|x\| u_0.$$

Clearly $E = \bigcup_{m,n=1}^{\infty} E_{m,n}$. This and the fact that E is of second category imply that at least one of the sets $E_{m,n}$ is not nowhere dense in E. Hence there exists a sphere $T(x_0, \rho)$: $\|x - x_0\| \le \rho$ such that some set $E_{m_0 n_0}$ is everywhere dense in $T(x_0, \rho)$ and $x_0 \in E_{m_0 n_0}$.

Consider the rings $T(x_0, r, \rho)$ and $T(\theta, r, \rho)$: $r \le \|x - x_0\| \le \rho$ and $r \le \|x\| \le \rho$ where $0 < r < \rho$.

Clearly $T(\theta, r, \rho) = T(x_0, r, \rho) \ominus x_0$ and for any $z = x - x_0 \in \Pi = (E_{m_0 n_0} \cap T(x_0, r, \rho)) \ominus x_0$ we have

$$-m_0(\|x\| + \|x_0\|) u_0 \leqslant A^{n_0}z = A^{n_0}x - A^{n_0}x_0 \leqslant (\|x\| + \|x_0\|) u_0.$$

Thus for any $z \in \Pi$

$$\beta \| z \| u_0 = - m_0 \frac{\rho + 2 \| x_0 \|}{r} \| z \| u_0 \leqslant A^{n_0} z \leqslant m_0 \frac{\rho + 2 \| x_0 \|}{r} \| z \| u_0 = \beta \| z \| u_0.$$

This and the density of Π in the ring $T(\theta, r, \rho)$ together with the continuity of A imply that for all $x \in T(\theta, r, \rho)$

$$- \beta \| x \| u_0 \leqslant A^{n_0} x \leqslant \beta \| x \| u_0. \tag{6}$$

Since along with any $x \in T(\theta, r, \rho)$, (6) holds also for any $y = tx$, $t \geq 0$, (6) is valid for all $x \in E$.

The theorem is proved.

5. Consider the case when K is a cone. For some $x \in E$ let

$$- \alpha u_0 \leqslant x \leqslant \alpha u_0. \tag{7}$$

The smallest $\alpha \geq 0$ for which (7) holds is called the u_0-*norm* ([3], [4]) of x: $\|x\|_{u_0}$. Note that the u_0-norm has all the usual properties of a norm. It is defined on some linear set $E_0 \subset E$ which may coincide with all of E.

Definition 1. An operator A is called u_0-*continuous* (see [4], [5]) *at* $x = x_0$ if

$$\lim_{\|x - x_0\| \to 0} \| Ax - Ax_0 \|_{u_0} = 0.$$

Definition 2. An operator A is called u_0 *continuous relative to the cone* K *at* $x_0 = x$ if

$$\lim_{x \geqslant x_0, \|x - x_0\| \to 0} \| Ax - Ax_0 \|_{u_0} = 0.$$

Clearly if K is a cone and A satisfies the conditions of Theorem 3 then A^{n_0} is u_0-continuous.

If A satisfies the conditions of Theorem 1 then A^{n_0} is u_0-continuous relative to the cone K.

6. Just as in [2] one proves the following assertions.

Theorem 4. *If the generating semigroup K is closed, then there exists a universal number $M > 0$ such that for any $x \in E$ there exists an element $u(x) \in K$ such that*

$$x \leqslant u(x),$$
$$\| u(x) \| \leqslant M \| x \|.$$

Theorem 5. *Let $K \subset E$ be a closed generating semigroup and K_1 a normal cone in E_1. Then each positive $(AK \subset K_1)$ additive operator A from E to E_1 is continuous.*

From Theorem 5 it follows in particular that if an additive functional f is positive on a closed generating semigroup K, then it is linear.

§ 2. Abelian classes of positive linear operators

1. **Theorem 6.** *Let K be a closed semigroup and $\Gamma = \{A\}$ a commutative class of positive linear operators such that:*

a) *the operators $A \in \Gamma$ have a common eigenvector $v \in K$ $(-v \notin K)$:*

$$v = Av;$$

b) *for any $A \in \Gamma$ and any $x \in K$ there exist a natural number n and a $\beta > 0$ such that*

$$A^n x \leqslant \beta v.$$

Then there exists a positive additive homogeneous functional ψ defined on the linear hull of K such that

$$\psi(Ax) = \psi(x)$$

for all $A \in \Gamma$.

Proof. Let $\|x\|_r = \inf t$, where the inf extends over all t such that

$$-tv \leqslant x \leqslant tv. \tag{8}$$

Let E_0 denote the set of all $x \in E$ where $\|x\|_v$ is defined. Clearly $\|x\|_v$, defined on E_0, is such that

$$\|x\|_v \geqslant 0,$$

$$\|\lambda x\|_v = |\lambda| \, \|x\|_v, \quad \|x + y\|_v \leqslant \|x\|_v + \|y\|_v.$$

Since $Av = v$, from (8) it follows that $-tv \leqslant Ax \leqslant tv$, i.e.

$$\|Ax\|_v \leqslant \|x\|_v. \tag{9}$$

Put the norm $\|x\|^* = \|x\| + \|x\|_v$ on E_0. Then E_0 becomes a Banach space and $K_0 = K \cap E_0$ is a closed solid semigroup. v is an interior point of K_0, since for all x in $\|x-v\|^* = \|x-v\| + \|x-v\|_v \leq 1$ we have

$$-v \leqslant x - v \leqslant v,$$

or

$$\theta \leqslant x \leqslant 2v, \quad \text{i.e. } x \in K_0.$$

Clearly the class $\Gamma = \{A\}$ leaves E_0 and K_0 invariant.

Consider the linear set G of elements of the form

$$y = A_1x_1 - x_1 + A_2x_2 - x_2 + \ldots + A_nx_n - x_n, \tag{10}$$

where

$$x_i \in E_0, \quad A_i \in \Gamma \quad (i = 1, \ldots, n;\ n = 1, 2, \ldots).$$

Let us show that G contains no interior points of K_0.

In fact, let some element of the form (10) lie in the interior of K_0. Then there is a $\rho > 0$ such that $y \gg \rho v$. Let N be a natural number. Set

$$B_N = \frac{1}{N^n} \sum_{k_1,\ldots,k_n=1}^{N} A_1^{k_1} A_2^{k_2} \ldots A_n^{k_n}. \tag{11}$$

Clearly $B_N v = v$. Hence

$$B_N y \gg \rho B_N v = \rho v, \quad \text{i.e. } \| B_N y \|_v \geqslant \rho \quad (-v \notin K_0). \tag{12}$$

On the other hand, by (9), $\|A_1^{k_1} A_2^{k_2} \cdots A_n^{k_n} x_j\|_v \leq \|x_j\|_v$ $(j = 1, \cdots, n)$ and the vector $N^n B_N y$, provided we use definitions (10) and (11), can be written in the form of a sum of $2N^{n-1}n$ monomials $A_1^{k_1} A_2^{k_2} \cdots A_n^{k_n} x_j$. Hence

$$\| B_N y \|_v \leqslant \frac{2n}{N} \max \{\| x_1 \|_v, \| x_2 \|_v, \ldots, \| x_n \|_v\}.$$

For sufficiently large N this contradicts (12).

Thus G does not meet the interior of K_0. Hence there exists a functional ϕ that is positive on K_0 (see [1], p. 12 (transl. p. 212), Corollary 1.2) such that $\phi(y) = 0$ $(y \in G)$.

In particular, setting $y = Ax - x$ $(x \in E_0,\ A \in \Gamma)$, we get

$$\varphi(Ax) = \varphi(x).$$

Let $A_0 \in \Gamma$ be any fixed operator. Since the conditions of Theorem 1 hold for A_0, it follows that for some fixed n_0 and $\beta > 0$, for each $x \in K$ we have $A_0^{n_0} x \leqslant \beta \|x\| \cdot v$, i.e. $A_0^{n_0} x \in K_0$.

Hence, for all $x \in K$

$$\varphi(AA_0^{n_0}x) = \varphi(A_0^{n_0}x) \geqslant 0.$$

Set

$$\varphi(A_0^{n_0}x) = \psi(x). \tag{13}$$

Then $\psi(Ax) = \psi(x)$ for any $A \in \Gamma$.

The theorem is proved.

2. **Theorem** 7. *Let the conditions of Theorem* 6 *be satisfied for a commutative class* $\Gamma = \{A\}$ *of positive linear operators, as well as at least one of the following conditions*:

1) *the semigroup* K *is generating,*

2) *the operators* $A \in \Gamma$ *are* v-*bounded on* E.

Then the class $\{A^*\}$ *of conjugate operators* A^* *has a common eigenvector* ψ *in the conjugate semigroup* K^* *of positive linear functionals* (*see* [1]):

$$A^*\psi = \psi.$$

Proof. First let us show that if either 1) or 2) holds then the functional in (13) is defined on all of E and is linear. If 1) holds this is clear, since the linear hull of K is E and an additive functional of ψ which is positive on the closed semigroup K is linear.

Now if 2) holds Theorem 3 implies that $A_0^{n_0} E \subset E_0$. Hence in this case also ψ is defined on all of E. Moreover, from Theorem 3 and the positivity of ψ it follows (see [5]) that $\lim_{\|x\| \to 0} \psi(x) = 0$. So the functional ψ is linear in any case. This and the equality $\psi(Ax) = \psi(x)$ imply that $A^*\psi = \psi$. The theorem is proved.

3. Now we give a theorem of Kreĭn (see [1], Theorem 3.3):

Let K be a solid semigroup. For any commutative class $\Gamma = \{A\}$ of linear operators A mapping the interior of K into itself, there is always a positive functional $\psi \in K^*$ which is a common eigenvector of all the conjugate operators:

$$A^*\psi = \lambda_A \psi \quad (\lambda_A > 0, \ A \in \Gamma).$$

With this theorem let us prove an assertion stronger than Theorems 6 and 7.

Theorem 8. *Let K be a closed semigroup and $\Gamma = \{A\}$ a commutative class of positive linear operators such that*

a) *for any $A \in \Gamma$ and $x \in K$ there exist a natural number n and a scalar $\gamma > 0$ such that $A^n x \leqslant \gamma u_0$, where u_0 is a fixed element of $K(-u_0 \notin K)$,*

b) *for any $A \in \Gamma$ there exist $\alpha, \beta > 0$ such that*

$$\alpha u_0 \leqslant A u_0 \leqslant \beta u_0.$$

Then there exists a positive additive homogeneous functional $\psi \neq 0$ defined on the linear hull of K such that

$$\psi(Ax) = \lambda_A \psi(x) \quad (\lambda_A > 0)$$

for all $A \in \Gamma$.

Proof. Just as in the proof of Theorem 6, consider the linear set E_0 consisting of all $x \in E$ where $\|x\|_{u_0}$ is defined:

$$\|x\|_{u_0} = \inf t,$$

where the inf is taken over all t such that $-tu_0 \leqslant x \leqslant tu_0$.

Introduce on E_0 the norm $\|x\|^* = \|x\| + \|x\|_{u_0}$ Then E_0 is a complete Banach space and $K_0 = K \cap E_0$ is a closed solid semigroup.

Clearly the class $\Gamma = \{A\}$ leaves E_0 and K_0 invariant. Hence each operator A, by condition b) of the theorem, maps the interior of K_0 into itself. Hence for the commutative class $\Gamma = \{A\}$ all the conditions of Kreĭn's theorem hold.

Hence there exists a positive linear functional $\phi \neq 0$ such that

$$\varphi(Ax) = \lambda_A \varphi(x) \quad (\lambda_A > 0, \ x \in E_0)$$

for all $A \in \Gamma$.

Let $A \in \Gamma$ be any fixed operator. Since all the conditions of Theorem 1 hold for A_0, it follows that for some fixed n_0 and $\beta_0 > 0$, for each $x \in K$ we have $A_0^{n_0} x \leqslant \beta_0 \|x\|_{u_0}$, i.e. $A_0^{n_0} x \in K$. Hence for all $x \in K$

$$\varphi(A A_0^{n_0} x) = \lambda_A \varphi(A_0^{n_0} x) \geqslant 0.$$

Set

$$\psi(x) = \varphi(A_0^{n_0} x). \tag{14}$$

Clearly ψ is a positive additive homogeneous functional and

$\psi(Ax) = \lambda_A \psi(x)$ for any $A \in \Gamma$.

The theorem is proved.

Corollary. *In the conditions of Theorem* 8 *let* b) *be replaced by the weaker condition*

b′) *there exists a natural number* $m_0 > 1$ *such that for all* $A \in \Gamma$

$$\alpha u_0 \leqslant A^{m_0} u_0 \leqslant \beta u_0,$$

where $\alpha = \alpha(A) > 0$, $\beta = \beta(A) > 0$,

Then the following assertions are valid.

1) *There exists a positive additive homogeneous functional* $\psi \neq 0$ *defined on the linear hull of* K *such that for all* $A \in \Gamma$

$$\psi(A^{m_0} x) = \lambda_A \psi(x) \quad (\lambda_A > 0).$$

2) *There exists a positive additive homogeneous functional* $g_A \neq 0$ *defined on the linear hull of* K *such that for all* $A \in \Gamma$

$$g_A(Ax) = \mu_A g_A(x) \quad (\mu_A > 0).$$

In fact, $\{A^{m_0}\}$ satisfies all the conditions of Theorem 8. Hence there exists a positive additive homogeneous functional $\psi \neq 0$ defined on the linear hull of K such that for all $A \in \Gamma$

$$\psi(A^{m_0} x) = \lambda_A \psi(x) \quad (\lambda_A > 0).$$

Let

$$\mu_A = \sqrt[m_0]{\lambda_A} \text{ and } y = \mu_A^{m_0-1} x + \mu_A^{m_0-2} Ax + \ldots + A^{m_0-1} x.$$

Clearly $\psi(Ay) = \mu_A \psi(y)$. Set

$$g_A(x) = \mu_A^{m_0-1} \psi(x) + \mu_A^{m_0-2} \psi(Ax) + \ldots + \psi(A^{m_0-1} x).$$

Then

$$g_A(Ax) = \mu_A g_A(x).$$

4. **Theorem** 9. *Let* K *be a cone and let* $\Gamma = \{A\}$ *be a commutative class of positive linear operators such that*:

a) *For each* $A \in \Gamma$ *and any* $x \in K$ *there exist a natural number* n *and a scalar* $\gamma > 0$ *such that*

$$A^n x \leqslant \gamma u_0,$$

where u_0 *is a fixed nonzero element in* K.

b) *There exists a fixed natural number* m_0 *such that for any* $x \in K$ $(x \neq \theta)$, *for all* $A \in \Gamma$,

$$A^{m_0}x \gg \alpha u_0,$$

where $\alpha = \alpha(A, x) > 0$.

Then the following assertions are valid.

1) *There exist a constant* p_0 *and a positive additive homogeneous functional* $\psi \neq 0$ *defined on the linear hull of* K *such that for all* $A \in \Gamma$

$$\psi(A^{p_0}x) = \lambda_A \psi(x) \quad (\lambda_A > 0).$$

2) *For any* $A \in \Gamma$ *there exists a positive additive homogeneous functional* $g_A \neq 0$ *defined on the linear hull of* K *such that*

$$g_A(Ax) = \mu_A g_A(x) \quad (\mu_A > 0).$$

Proof. Let $A_0 \in \Gamma$ be a fixed operator. Since A_0 satisfies the conditions of Theorem 1 there exist a fixed natural number p and a scalar $\beta > 0$ such that $A_0^p x \leqslant \beta \|x\| u_0$ for all $x \in K$.

Let us write $p_0 = \max\{m_0, p\}$. Then for any nonzero $x \in K$, for some $\alpha_1 > 0$ and $\beta_1 > 0$ we have

$$\alpha_1 u_0 \leqslant A_0^{m_0}(A_0^{p_0 - m_0}x) = A_0^{p_0}x = A_0^p(A_0^{p_0 - p}x) \leqslant \beta_1 u_0.$$

Let E_0 be the set of $y \in E$ where the u_0-norm $\|x\|_{u_0}$ is defined, and let E_1 be the set of $y \in E_0$ where $A_0^{p_0} y = \theta$. Clearly E_1 either consists only of θ or contains a nonzero element.

Suppose first that E_1 consists only of θ. Set $E_2 = A_0^{p_0} E_0$. Introduce on E_2 the norm $\|z\|^*$ $(z = A_0^{p_0} y,\ y \in E_0)$, setting

$$\|z\|^* = \|y\| + \|y\|_{u_0} + \|z\|_{u_0}.$$

Then E_2 becomes a Banach space with solid cone $K_2 = K_0 \cap E_2$ $(K_0 = K \cap E_0)$.

In fact, for u_0, for some $\alpha_2 > 0$, $\beta_2 > 0$ we have

$$\alpha_2 u_0 \leqslant z = A^{p_0} u_0 \leqslant \beta_2 u_0.$$

Hence for all $z = A^{p_0} y$ $(y \in E_0)$ in the sphere

$$\|z - z_0\|^* = \|y - u_0\| + \|y - u_0\|_{u_0} + \|z - z_0\|_{u_0} \leqslant \min\{1, \alpha_2\}$$

we get

$$-u_0 \leqslant y - u_0, \quad -\alpha_2 u_0 \leqslant z - z_0,$$

$$y \geqslant \theta; \quad z \geqslant z_0 - \alpha_2 u_0 = A^{p_0} u_0 - \alpha_2 u_0 \geqslant \theta.$$

The operators A^{p_0} $(A \in \Gamma)$ act on E_2 and map the interior of K_2 into itself. This and the commutativity of $\Gamma = \{A\}$ imply that the A^{p_0} satisfy all the conditions of the above theorem of Kreĭn. Hence

1) there exists a positive additive homogeneous functional $\phi \neq 0$ defined on the linear hull of K such that for all $A \in \Gamma$

$$\varphi(A^{p_0} z) = \lambda_A \varphi(z) \quad (\lambda_A > 0);$$

2) there exists a positive additive homogeneous functional $f_A \neq 0$ defined on the linear hull of K_0 such that for all $A \in \Gamma$

$$f_A(Az) = \mu_A f_A(z) \quad (\mu_A > 0). \tag{15}$$

Since $A_0^{p_0} K \subset K_0$ and $z = A_0^{p_0} y = A_0^{2p_0} x$, the functionals

$$\psi(x) = \varphi(A_0^{2p_0} x), \qquad g_A(x) = f_A(A_0^{2p_0} x)$$

will be the desired ones.

Now suppose that E_1 contains nonzero elements. Clearly E_0 and E_1 are complete in the norm $\|y\| + \|y\|_{u_0}$.

Consider the factor space $F = E_0 / E_1$ of classes $Y = y \oplus E_1$ $(y \in E_0)$ with the norm

$$\| Y \| = \inf_{y \in Y} (\| y \| + \| y \|_{u_0}).$$

As is known, the space F is complete. Set $E_2 = A_0^{p_0} E_0$. Introduce on E_2 the norm

$$\| z \|^* = \| Y \| + \| z \|_{u_0},$$

where $z = A_0^{p_0} y$, $y \in E_0$, and the class Y contains y. Then E_2 becomes a Banach space with solid cone $K_2 = K_0 \cap E_2$. The operators A^{p_0} $(A \in \Gamma)$ act on E_2 and map the interior of K_2 into itself. Hence the theorem of Kreĭn is applicable in this case also. Hence the desired functionals ψ and g_A do exist.

The theorem is proved.

5. **Theorem** 10. *For the commutative class* $\Gamma = \{A\}$ *of positive linear operators let the conditions of Theorem* 8 (*Theorem* 9) *hold and also one of the following conditions*:

1) *the semigroup (cone)* K *is generating*,

2) *the operators* $A \in \Gamma$ *are* u_0*-bounded on* E.

Then the class $\{A^*\}$ $(\{A^{*p_0}\})$ *of conjugate operators* A^* (A^{*p_0}) *has a common eigenvector* ψ *in the conjugate semigroup* K^* *of positive linear functionals*:

$$A^*\psi = \lambda_A \psi \quad (A^{*p_0}\psi = \lambda_A \psi), \quad \lambda_A > 0.$$

Proof. Indeed, if one of the conditions 1) and 2) holds then the functional ψ (see (14) and (15)) is defined on all of E and is linear (see the proof of Theorem 7). Hence, because of

$$\psi(Ax) = \lambda_A \psi(x) \qquad (\psi(A^{p_0}x) = \lambda_A \psi(x)),$$

for all $A \in \Gamma$ we have

$$A^*\psi = \lambda_A\psi \quad (A^{*p_0}\psi = \lambda_A\psi).$$

The theorem is proved.

6. Using Theorem 10, let us prove the existence of an eigenvector for a positive linear operator A acting on a reflexive space E.

Theorem 11. *Let* E *be a reflexive space* $(E^{**} = E)$ *and let* $K \subset E$ *be a normal cone. Then any positive linear operator* A *has an eigenvector in the cone* K *provided the conjugate operator* A^* *is g-positive, i.e. for any* $f \in K^*$ $(f \neq 0)$ *there exist a natural number* n *and scalars* $\alpha, \beta > 0$ *such that*

$$\alpha g \leqslant A^{*n}f \leqslant \beta g,$$

where $g \in K^*$ $(g \neq 0)$ *is a fixed functional.*

Proof. Since the cone K is normal, the semigroup K^* (see [6]) is generating.

Hence for A^* all the conditions of Theorem 10 hold. Therefore A^{**} has an eigenvector z in K^{**}:

$$A^{**}z = \lambda_A z \quad (\lambda_A > 0).$$

But from the reflexivity of E it follows that $A^{**} = A$ and $K^{**} = K$. Hence $Az = \lambda_A z\,(\lambda_A > 0)$ for $z \in K$.

The theorem is proved.

BIBLIOGRAPHY

[1] M. G. Kreĭn and M. A. Rutman, *Linear operators leaving invariant a cone in a Banach space,* Uspehi Mat. Nauk 3 (1948), no. 1 (23), 3–95; English transl., Amer. Math. Soc. Transl. (1) 10 (1962), 199–325. MR 10,256.

[2] I. A. Bahtin, M. A. Krasnosel'skiĭ and V. Ja. Stecenko, *On the continuity of positive linear operators,* Sibirsk. Mat. Ž. 3 (1962), 156–160. (Russian) MR 25 #2451.

[3] M. A. Krasnosel'skiĭ, *Positive solutions of operator equations,* Fizmatgiz, Moscow, 1962; English transl., Noordhoff, Groningen, 1964. MR 26 #2862.

[4] L. V. Kantorovič, B. Z. Vulih and A. G. Pinsker, *Functional analysis and partially ordered spaces,* GITTL, Moscow, 1950. (Russian) MR 12,340.

[5] I. A. Bahtin, *Non-linear equations with concave and uniformly concave operators,* Dokl. Akad. Nauk SSSR 126 (1959), 9–12. (Russian) MR 21 #4366.

[6] M. Krein, *Propriétés fondamentales des ensembles coniques normaux dans l' espace de Banach,* C. R. (Dokl.) Acad. Sci. URSS 28 (1940), 13–17. MR 2,315.

Translated by:
J. Ellis

Amer. Math. Transl.
(2) Vol. 90, 1970

ON CONTINUOUS BRANCHES OF SEMI-EIGENVECTORS OF NONLINEAR OPERATORS*

UDC 513.882

I. A. BAHTIN

In this article we show that under certain conditions the solutions of systems of equations of the form

$$\begin{aligned} x &= \mu A(x, y), \\ y &= B(x, y) \end{aligned} \tag{1}$$

constitute continuous braches. The results are applied to the study of the distribution of the flow of neutrons and of temperature by volume of a reactor.

§ 1. Theorems on the existence of semi-eigenvectors

1. We will introduce some definitions and notation which we will use later.

Let E be a real Banach space partially ordered by the cone $K \subset E$; we write $x \leq y$ if $y - x \in K$ (cf. [1]).

Let A be some operator acting in E. A is said to be positive (cf. [2]) if $AK \subset K$. A is said to be monotone (cf. [2]) if for any pair $x, y \in K$, $x \leq y$ implies $Ax \leq Ay$.

A positive operator A is said to be u_0-positive (cf. [3]) if for any $x \in K$ $(x \neq \theta)$ there exist a natural number n and numbers $\alpha, \beta > 0$ such that

$$\alpha u_0 \leqslant A^n x \leqslant \beta u_0,$$

where u_0 is some fixed nonzero element of the cone K.

A positive monotone operator B is said to be a monotone minorant (cf. [2]) of the operator A on $K_r = K \cap T_r$, where T_r is the ball $\|x\| \leq r$, if $Ax \geq Bx$ $(x \in K_r)$ and if for some $u \in K$ $(u \neq \theta)$

$$cBtu \geqslant tu \quad (c > 0, \quad 0 \leqslant t < \gamma),$$

where γ is a number such that $x - \gamma u \notin K$ for all $x \in K_r$.

We will say that the set N of eigenvectors z of operator A: $Ax = \lambda x$ forms a continuous branch of infinite length (cf. [2]) if the boundary Γ of any open bounded set G containing the zero element θ has a nonnull intersection with N.

---

* Translation of Izv. Akad. Nauk SSSR Ser. Mat. 30 (1966), 1017–1026.

Henceforth we will use the following notational conventions.

Let E_1 and E_2 be two real Banach spaces with cones K_1 and K_2. Let $E = E_1 \times E_2$ be the Banach space of elements $z = (x, y)$ $(x \in E_1, y \in E_2)$ with norm $\|z\| = \|x\| + \|y\|$. Obivously $K = K_1 \times K_2$ is a cone in E.

Let A and B be operators mapping E into E_1 and E_2 respectively. Then the operator $C = (A, B)$ will operate in the space E. Let $C(\mu)$ denote the operator $(\mu A, B)$. It is easy to see that we can write system (1) in the form of one equation

$$z = C(\mu)z. \tag{2}$$

Let A_1, A_2 and B_1, B_2 represent respectively the operators A and B on the subspaces E_1 and E_2.

We will call the solution $z = (x, y)$ of equation (2) a semi-eigenvector of the operator C if $x \neq \theta$. The number μ $(\lambda = 1/\mu)$ will be called a semi-eigenvalue of operator C. The set of all semi-eigenvalues of C will be called the semispectrum of C. The set S of all semi-eigenvalues λ of C corresponding to semi-eigenvecters $z \in K$ will be called the positive semi-spectrum of C.

2. In this section we will present the central theorem of the article, an analog of the topological principle of M. A. Krasnosel'skiĭ (cf. [2], Chapter V, Theorem 1.1).

Theorem 1. *Let the positive operator $C = (A, B)$ be completely continuous. Let*

$$\inf_{z \in \Gamma \cap K} \| Az \| = a > 0, \tag{3}$$

where Γ is the boundary of some open bounded set G containing the zero element θ. Finally let all eigenvalues of B_2 corresponding to eigenvectors $y \in \Gamma \cap K_2$ be less than unity. Then C has at least one semi-eigenvector $z \in \Gamma$ in the cone K.

Proof. Suppose the contrary. Let C not have semi-eigenvectors on $\Gamma \cap K$. We will consider two cases: when $E = E_1 \times E_2$ is finite dimensional and when E is infinite dimensional.

First suppose that E is finite dimensional. We will construct an extension $\tilde{A}$ of A to $\overline{G}$ such that

$$\tilde{A}\overline{G} \subset K_1 \tag{4}$$

and

$$\inf_{z \in \overline{G}} \|\tilde{A}z\| > 0. \tag{5}$$

For the construction of $\tilde{A}$ we will consider some system of coordinates in the linear set L of minimal dimension containing $A(\Gamma \cap K)$. Let n be the dimension of L. Then on $\Gamma \cap K$ the operator A is equivalent to n continuous functions $\phi_1(z), \cdots, \phi_n(z)$. Extending them to $\overline{G}$ by Urysohn's Lemma we get the continuous extension $\overline{A}$ of A to $\overline{G}$, and $\overline{A}\,\overline{G} \subset L$.

Let Ω be the convex hull of the set $A(\Gamma \cap K)$. It is easy to see that $\Omega \subset K_1$ and that the distance $\rho(\Omega, \theta) > 0$. We will choose some interior point u in Ω, and we will let Φ be an operator coinciding with the identity operator I on Ω and mapping each point $x \in L$ $(x \in \Omega)$ into a point of the intersection of the segment connecting points u and x with the boundary of the set Ω. Φ is continuous.

Clearly $\widetilde{A} = \Phi \overline{A}$ satisfies conditions (4) and (5).

By an analogous method we will construct a continuous extension $\widetilde{B}$ of B to $\overline{G}$ such that $\widetilde{B}\,\overline{G} \subset K_2$.

It is easy to see that operator $\widetilde{C} = (\widetilde{A}, \widetilde{B})$ is positive on $\overline{G}$, and $\widetilde{C} = C$ for $z \in \Gamma \cap K$. Consequently $\widetilde{C}$ also does not have semi-eigenvectors on $\Gamma \cap K$.

We will consider the vector field

$$\Phi_t = I - (t\widetilde{A}, \widetilde{B}).$$

on the boundary Γ of G.

We will show that for sufficiently large $t > 0$ this field emits any fixed direction $u \in K_1$ $(u \neq \theta)$.

Suppose the contrary. Let there exist $u_1 \in K_1$ $(u_1 \neq \theta)$ and sequences $t_k \to \infty$ and $z_k = (x_k, y_k) \in \Gamma$ such that

$$\begin{cases} x_k - t_k\widetilde{A}(x_k, y_k) = \lambda_k u_1 & (\lambda_k > 0), \\ y_k - \widetilde{B}(x_k, y_k) = \theta. \end{cases}$$

We will choose a subsequence z_{k_n} such that the sequence $\widetilde{A}z_{k_n}$ converges to some u_2 $(u_2 \in K_1, u_2 \neq \theta)$. Then

$$A(x_{k_n}, y_{k_n}) - \frac{x_{k_n}}{t_{k_n}} \Rightarrow u_2.$$

Since

$$A(x_{k_n}, y_{k_n}) - \frac{x_{k_n}}{t_{k_n}} = -\frac{\lambda_{k_n}}{t_{k_n}} \cdot u_1,$$

we get $-u_1 \in K_1$ by passing to the limit. But this is impossible because $u_1 \in K_1$ and $u_1 \neq \theta$.

It is not hard to see that for any $t \geq 0$ the field Φ_t on Γ does not vanish. Consequently the rotation of these fields (cf. [2]) on Γ is equal. Let us calculate this number. On the one hand it is equal to zero, since we proved that for sufficiently large t the field Φ_t emits certain directions. On the other hand, the absolute value of this number equals unity, because the field $\Phi_0 = I - (0, B)$ is homotopic (cf. [2]) to the field I on Γ by virtue of the last condition of the theorem.

This contradiction completes the proof of the theorem for the finite-dimensional case.

Let us suppose now that E is infinite dimensional.

We will show that there a number $b > 0$ such that for all $z = (x, y) \in \Gamma \cap K$ and $t \geq 0$

$$\| A(x, y) - tx \| + \| B(x, y) - y \| \geqslant b. \tag{6}$$

In fact, if this were not so there would exist two sequences $z_k = (x_k, y_k) \in \Gamma \cap K$ and $t_k > 0$ such that

$$\| A(x_k, y_k) - t_k x_k \| < 1/k$$

and

$$\| y_k - B(x_k, y_k) \| < 1/k. \tag{7}$$

The numbers t_k satisfy the inequalities

$$\frac{1}{\| x_k \|}\left(\| A(x_k, y_k) \| - \frac{1}{k}\right) < t_k < \frac{1}{\| x_k \|}\left(\| A(x_k, y_k) \| + \frac{1}{k}\right).$$

By the conditions of the theorem the eigenvalues of B_2 corresponding to eigenvectors $y \in \Gamma \cap K_2$ are less than unity. Therefore by (7) the sequence $\|x_k\|$ is bounded from below by a positive number.

Consequently, since the sequences $\|x_k\|$ and $\|A(x_k, y_k)\|$ are uniformly bounded from above and from below by positive numbers, there exists a sequence of indices k_n such that the numbers t_{k_n} converge to some $t_0 \neq 0$ and the vectors $A(x_{k_n}, y_{k_n})$ and $B(x_{k_n}, y_{k_n})$ converge to some vectors $u \neq \theta$ and v. Then $y_{k_n} \Rightarrow v$ and $x_{k_n} \Rightarrow t_0^{-1}u$, because

$$\left\| x_{k_n} - \frac{u}{t_0} \right\| \leqslant \left\| x_{k_n} - \frac{A(x_{k_n}, y_{k_n})}{t_{k_n}} \right\|$$
$$+ \| A(x_{k_n}, y_{k_n}) \| \cdot \left| \frac{1}{t_{k_n}} - \frac{1}{t_0} \right| + \frac{1}{t_0} \| A(x_{k_n}, y_{k_n}) - u \|.$$

The vector $(t_0^{-1}u, v) \in \Gamma \cap K$ is semi-eigenvector of C. Thus we arrive at a contradiction.

Denote by A' and B' finite-dimensional operators approximating A and B on $\Gamma \cap K$ with accuracy up to $\delta = \min\{a/4, b/4\}$:

$$\| Az' - Az \| < \delta, \quad \| B'z - Bz \| < \delta \quad (z \in \Gamma \cap K) \tag{8}$$

and simultaneously satisfying the conditions

$$A'(\Gamma \cap K) \subset K_1, \quad B'(\Gamma \cap K) \subset K_2.$$

The possibility of such an approximation follows immediatley, for example, from the Schauder construction of approximating finite-dimensional operators (cf. [2], Chapter II, § 3.2).

We will let E_1' and E_2' denote finite-dimensional subspaces containing $A'(\Gamma \cap K)$ and $B'(\Gamma \cap K)$ respectively. Let K_1' and K_2' be the cones $K_1 \cap E_1'$ and $K_2 \cap E_2'$. Obviously $E' = E_1' \times E_2'$ is a finite-dimensional space and $G' = G \cap E'$ and $K' = K_1' \times K_2'$ are respectively an open set and a cone in this space. Let $\Gamma' \subset \Gamma$ be the boundary of the set G'. From (3), (6) and (8) it follows that

$$\| A'(x, y) - tx \| + \| B'(x, y) - y \| \geqslant \| A(x, y) - tx \|$$
$$- \| A'(x, y) - A(x, y) \| + \| B(x, y) - y \| - \| B'(x, y) - B(x, y) \| > \frac{b}{2}$$
$$(z = (x, y) \in \Gamma' \cap K'; \quad t \geqslant 0)$$

and

$$\inf_{z \in \Gamma' \cap K'} \| A'z \| > \frac{3}{4} a.$$

Thus the operator $C' = (A', B')$ does not have semi-eigenvectors on $\Gamma' \cap K'$.

On the other hand, C' has semi-eigenvectors on $\Gamma' \cap K'$, since all the conditions of the theorem are fulfilled for C' in the finite-dimensional space E', and for a finite-dimensional space the existence of semi-eigenvectors is already established.

So the assumption that $C = (A, B)$ does not have semi-eigenvectors on $\Gamma \cap K$ leads to a contradiction. The theorem is proved.

We point out that the assertion of Theorem 1 is still valid if the last condition is altered in the following way: there exists a completely continuous extension $\widetilde{B}$ of B from $\Gamma \cap K$ onto $\overline{G}$ such that

a) $\widetilde{B}z \in K_2$ for all $z \in \overline{G}$;

b) the field $y - \widetilde{B}(\theta, y)$ on $\Gamma_2 = \Gamma \cap E_2$ does not have null vectors;

c) on Γ_2 the rotation of the vector field $y - \widetilde{B}(\theta, y)$ is different from zero.

3. In applications it is convenient to make use of the following theorem.

Theorem 2. *Let the positive operator* $C = (A, B)$ *be completely continuous. For all* $y \in (K_2)_r = K_2 \cap T_r$ *let the operators* $A_y x = A(x, y)$ *on* $(K_1)_r = K_1 \cap T_r$ *have a common minorant* A_0. *Finally let all eigenvalues of operator* B_2 *corresponding to eigenvectors* $y \in (K_2)_r$ *be less than unity. Then the semi-eigenvectors of* C *form a continuous branch of length* r *in the cone* K.

Proof. Since the operator A_0 is a monotone minorant on $(K_1)_r$ for all operators A_y $(y \in (K_2)_r)$, we have that

a) A_0 is monotone;

b) $A_y x \geq A_0 x$ $(y \in (K_2)_r)$ for all $x \in (K_1)_r$;

c) there exists an element $u \in K_1$ $(u \neq \theta)$ such that

$$cA_0 tu \geqslant tu \qquad (c > 0, \quad 0 \leqslant t < \gamma),$$

where γ is a number such that $x - \gamma u \notin K_1$ for all $x \in (K_1)_r$.

Let Γ be the boundary of any open set $G \subset T_r$ containing θ. Clearly for sufficiently large n the operators

$$C_n(x, y) = \left(A\left(x + \frac{u}{n}, y\right), B(x, y)\right)$$

on $\Gamma \cap K$ satisfy all conditions of Theorem 1. Consequently for such n the operators C_n have semi-eigenvectors $z_n = (x_n, y_n)$ on $\Gamma \cap K$:

$$\begin{cases} x_n = \mu_n A\left(x_n + \dfrac{u}{n}, y_n\right), \\ y_n = B(x_n, y_n). \end{cases}$$

We will show that $\mu_n \leq c$. Suppose the contrary and let $\mu_{n_0} > c$ for some n_0. We denote by t_0 the maximal number for which $x_{n_0} - t_0 u \geq \theta$. Obviously $t_0 \in [0, \gamma)$. Hence

$$x_{n_0} - t_0 u = \mu_{n_0} A\left(x_{n_0} + \frac{u}{n_0}, y_{n_0}\right) - t_0 u \geqslant \mu_{n_0} A_0\left(x_{n_0} + \frac{u}{n_0}\right) - cA_0 t_0 u$$
$$\geqslant (\mu_{n_0} - c) A_0\left(x_{n_0} + \frac{u}{n_0}\right) \geqslant (\mu_{n_0} - c) A_0 \frac{u}{n_0} \geqslant \frac{\mu_{n_0} - c}{c} \cdot \frac{u}{n_0},$$

that is,

$$x_{n_0} - \left(t_0 + \frac{\mu_{n_0} - c}{cn_0}\right) u \geqslant \theta,$$

which contradicts the maximality of t_0.

Because the sequence μ_n is bounded and C is completely continuous, it follows that there exists a subsequence of indices n_k for which

$$\mu_{n_k} \to \mu_0 \qquad (\mu_0 \neq 0),$$
$$A\left(x_{n_k} + \frac{u}{n_k},\ y_{n_k}\right) \Rightarrow \frac{x_0}{\mu_0}, \quad B(x_{n_k}, y_{n_k}) \Rightarrow y_0.$$

Thus

$$z_{n_k} = (x_{n_k}, y_{n_k}) \Rightarrow (x_0, y_0) = z_0 \in \Gamma \cap K,$$

and

$$z_0 = C(\mu_0) z_0.$$

The inequality $x_0 \neq \theta$ follows from the last condition of the theorem, as does $\mu_0 \neq 0$. Thus $z_0 \in \Gamma \cap K$ is a semi-eigenvector of C. The theorem is proved.

§ 2. Critical condition of a reactor with a nonlinear relation between temperature and the flow of neutrons

1. Let the distribution of neutron flow Φ and the temperature T relative to the volume V of a reactor be described (cf. [4]) by the equations

$$\begin{cases} \nabla^2\Phi + \mu a(T)\Phi = 0, & \Phi|_\Sigma = 0; \\ \nabla^2 T + b(T)\Phi = 0, & T|_\Sigma = T_0, \end{cases}$$

where Σ is the boundary of the reactor, which is assumed to be sufficiently smooth and convex. The coefficients $a(T)$ and $b(T)$ are positive and continuous in the physical sense (cf. [4]).

By changing variables the boundary conditions can be made homogenous, after which the system takes the form

$$\begin{cases} \nabla^2\Phi + \mu a(T + T_0)\Phi = 0, & \Phi|_\Sigma = 0; \\ \nabla^2 T + b(T + T_0)\Phi = 0 & T|_\Sigma = 0. \end{cases} \tag{9}$$

By means of the Green's function $\mathcal{K}(P, Q)$ it is possible to convert this system into an equivalent system of integral equations:

$$\left|\begin{array}{l} \Phi(P) = \mu \int\limits_V \mathcal{K}(P, Q) a[T(Q) + T_0]\Phi(Q)\,dQ = \mu A(\Phi, T), \\ \\ T(P) = \int\limits_V \mathcal{K}(P, Q) b[T(Q) + T_0]\Phi(Q)\,dQ = B(\Phi, T). \end{array}\right. \tag{10}$$

O. B. Moskalev studied system (10) in [4]. We will study it under different constraints.

In the following the roles of E_1, E_2 and K_1, K_2 will be played respectively by the space $C(V)$ of functions continuous on V and the cone of non-negative functions of $C(V)$.

In [4] it is shown that $C = (A, B)$ is completely continuous and that there exists an element $w_0 = (u_0, v_0) \in K$ such that for any $z = (\Phi, T) \in K$ $(\Phi \neq \theta, T \neq \theta)$

$$\alpha w_0 \leqslant Cz \leqslant \beta w_0,$$

where $\alpha = \alpha(z) > 0$, $\beta = \beta(z) > 0$.

2. **Theorem 3.** *The semi-eigenvectors* $z = (\Phi, T) \in K$ *of system* (10) *form a continuous branch of infinite length.*

Proof. We will investigate the set K_r where r is an arbitrary positive number. Let

$$a_r = \min_{0 \leqslant T \leqslant r} a(T + T_0).$$

Obviously the operator

$$A_r \Phi(P) = a_r \int_V \mathcal{K}(P, Q) \Phi(Q) dQ$$

is a common monotone minorant of the operators $A_T \Phi = A(\Phi, T)$ $(T \in (K_2)_r)$ on $(K_1)_r$. Since $B(\theta, T) = \theta$, the spectrum of the operator B_2 consists of one number, zero.

Thus for $C = (A, B)$ all the conditions of Theorem 2 are fulfilled. Hence, since r was chosen arbitrarily, the semi-eigenvectors of system (10) form a continuous branch of infinite length in the cone K. The theorem is proved.

3. We will now prove that under certain conditions the positive semi-spectrum of system (10) contains some interval.

For this we will need the following analog of Theorem 5.9 of M. A. Krasnosel'skiĭ [3].

Theorem. *Let some set* N *of semi-eigenvectors of the completely continuous operator* C *form a closed continuous branch of infinite length. Let the eigenvalues of* C *tend to the limits* λ_θ *and* λ_∞ *when the norms of the semi-eigenvectors from branch* N *tend respectively to zero and to infinity. Finally, for the branch* N *suppose that* $\|z\| \to 0$ *and* $\|x\| \to \infty$ *follow respectively from* $\|x\| \to 0$ *and* $\|z\| \to \infty$. *Then the semispectrum of* C *contains the whole interval* $(\lambda_\theta, \lambda_\infty)$ *besides, perhaps, the point zero.*

Theorem 4. *Let the functions* $a(T + T_0)$ *and* $b(T + T_0)$ *of system* (10) *have the following properties:*

1) *The limit*

$$\lim_{T \to \infty} a(T + T_0) = a_0 < \infty \tag{11}$$

exists.

2. *For all* $T \geq 0$

$$0 < b_1 \leqslant b(T + T_0) \leqslant \alpha_0 + \beta_0 T^{q_0}, \tag{12}$$

where α_0, $\beta_0 > 0$ *and* $q_0 \in (0, 1)$.

Then the positive semispectrum S *of* (10) *completely fills the interval* $(\lambda_0, \alpha_0 \lambda_0 / \alpha(T_0))$, *where* λ_0 *is an eigenvalue of the operator*

$$A_0 \Phi(P) = a(T_0) \int_V \mathcal{K}(P, Q) \Phi(Q) \, dQ,$$

corresponding to the positive eigenvector $\Phi_0(P)$.

Proof. We will show that the theorem formulated above is applicable to system (10). The proof of Theorem 4 will then follow.

By Theorem 3 the semi-eigenvectors $z = (\Phi, T) \in K$ of (10) form a closed continuous branch N of infinite length. Obviously

$$\|T\| \leqslant (\alpha_0 + \beta_0 \|T\|^{q_0}) \|\Phi\| \cdot \max \int_V \mathcal{K}(P, Q) \, dQ,$$

if $z = (\Phi, T) \in N$. Thus for semi-eigenvectors $z = (\Phi, T) \in N$ the convergence $\|\Phi\| \to 0$ implies $\|z\| \to 0$ and $\|z\| \to \infty$ implies $\|\Phi\| \to \infty$.

Finally we will show that for semi-eigenvectors $z = (\Phi, T) \in K$:

$$\begin{cases} \lambda \Phi(P) = \int_V \mathcal{K}(P, Q) a[T(Q) + T_0] \Phi(Q) \, dQ, \\ T(P) = \int_V \mathcal{K}(P, Q) b[T(Q) + T_0] \Phi(Q) \, dQ, \end{cases}$$

$\|z\| \to 0$ implies $\lambda \to \lambda_0$ and $\|z\| \to \infty$ implies $\lambda \to \alpha_0 \lambda_0 / \alpha(T_0)$.

Let $\|\Phi_n\| \to 0$. Then since $\|T_n\| \to 0$, the sequence λ_n:

$$\lambda_n \frac{\Phi_n(P)}{\|\Phi_n\|} = \int_V \mathcal{K}(P, Q) \, a[T_n(Q) + T_0] \frac{\Phi_n(Q)}{\|\Phi_n\|} dQ,$$

will be bounded (cf. [2]) from above and from below by positive numbers. Therefore there exists a subsequence of numbers n_k such that

$$\lambda_{n_k} \to \bar{\lambda} \neq 0, \qquad \frac{\Phi_n}{\|\Phi_n\|} \Rightarrow \bar{\Phi}.$$

Obviously

$$\bar{\lambda}\bar{\Phi}(P) = a(T_0)\int_V \mathcal{K}(P, Q)\bar{\Phi}(Q)\,dQ.$$

Hence since (12) is u_0-positive, it follows that $\bar{\lambda} = \lambda_0$.

Thus $\lambda_n \longrightarrow \lambda_0$.

We will now assume that $\|\Phi\| \longrightarrow \infty$. We consider two cases: a) $a_0 = 0$, b) $a_0 \neq 0$.

In case a) necessarily $\lambda \longrightarrow 0$. Let us suppose the contrary. Without loss of generality we can assume that

$$\frac{\Phi}{\|\Phi\|} \Rightarrow \Phi_0 \neq \theta.$$

From this and from the relations

$$\frac{T(P)}{\|\Phi\|} = \int_V \mathcal{K}(P, Q)\, b\,[T(Q) + T_0]\,\frac{\Phi(Q)}{\|\Phi\|}\,dQ$$

$$\geqslant b(T_0)\int_V \mathcal{K}(P,Q)\frac{\Phi(Q)}{\|\Phi\|}\,dQ \Rightarrow b(T_0)\int_V \mathcal{K}(P, Q)\Phi_0(Q)\,dQ$$

if follows that

$$T \geqslant \alpha_\Phi \cdot v_0,$$

where $\alpha_\Phi \longrightarrow \infty$ when $\|\Phi\| \longrightarrow \infty$. But then by property 1) and

$$\lambda\,\frac{\Phi(P)}{\|\Phi\|} = \int_V \mathcal{K}(P, Q)\, a\,[T(Q) + T_0]\,\frac{\Phi(Q)}{\|\Phi\|}\,dQ, \tag{13}$$

the number $\lambda \longrightarrow 0$, and we have a contradiction.

We will now assume that $0 < a_0 < \infty$. By continuity of the positive function $a(T + T_0)$ there exist numbers $a_1, a_2 > 0$ such that for all $T \geq 0$

$$a_1 \leqslant a(T + T_0) \leqslant a_2.$$

From this and (13) it follows that the semi-eigenvalues λ are bounded from above and from below by positive numbers. Therefore, as in case a), we can show that

$$T \geqslant \alpha_\Phi \cdot v_0,$$

where $\alpha_\Phi \longrightarrow \infty$ when $\|\Phi\| \longrightarrow \infty$.

Thus by property 1) and (13)

$$\lambda \longrightarrow \frac{a_0}{a(T_0)} \cdot \lambda_0.$$

Thus we can apply to system (10) the previously mentioned theorem with constants

$$\lambda_\theta = \lambda_0, \quad \lambda_\infty = \frac{a_0}{a(T_0)} \cdot \lambda_0.$$

Thus the theorem is proved.

BIBLIOGRAPHY

[1] M. G. Kreĭn and M. A. Rutman, *Linear operators leaving invariant a cone in a Banach space,* Uspehi Mat. Nauk 3 (1948), no. 1 (23), 3–95; English transl., Amer. Math. Soc. Transl. (1) 10 (1962), 199–325. MR 10,256.

[2] M. A. Krasnosel'skiĭ, *Topological methods in the theory of nonlinear integral equations,* GITTL, Moscow, 1956; English transl., Macmillan, New York, 1964. MR 20 #3464; 28 #2414.

[3] ———, *Positve solutions of operator equations,* Fizmatgiz, Moscow, 1962; English transl., Noordhoff, Groningen, 1964. MR 26 #2862.

[4] O. B. Moskalev, *Critical condition of a reactor with a non-linear relation between temperature and neutron flow.* Ž. Vyčisl. Mat. i Mat. Fiz. 3 (1963) 327–336 = USSR Comput, Math. and Math. Phys. 3 (1963), 430–441.

Translated by:
H. N. Parker